D1683239

Prairie Plants of Northern Illinois: Identification and Ecology

By Russell R. Kirt
College of DuPage

Illustrations by Henrietta H. Tweedie and Roberta L. Simonds

Copyright © 1989
STIPES PUBLISHING COMPANY

Second Printing 1991

ISBN 0-87563-340-4

Published By
STIPES PUBLISHING COMPANY
10-12 Chester Street
Champaign, IL 61820

ACKNOWLEDGEMENTS

Encouragement for writing this book came from many persons: notably my wife, Pamela, and College of DuPage students. Much enthusiasm and all the illustrations were provided by Henrietta H. Tweedie and Roberta L. Simmonds.

David Malek and Lynn Fancher from College of DuPage reviewed and proofread the manuscript. Their comments were helpful and appreciated.

My most sincere gratitude goes to Ray Schulenberg who not only taught and inspired me about "prairie" but also gave me information, advice, and constructive criticism in the preparation of this book.

Some prairie flowers were reprinted from *Wildflowers of the Great Lakes Region* by permission of Chicago Review Press. Cover illustration by Henrietta H. Tweedie.

No illustrations in this book may be used without permission in writing from the illustrator(s).

INTRODUCTION

This book is written and illustrated to aid the novice in the identification of 63 prairie plants native to northern Illinois and currently thriving in College of DuPage's Restoration Prairies.

In this book, the plants are arranged by family because of the prominence of three families in the prairie, i.e., the grass family (Poaceae), the legume family (Fabaceae) and the composite family (Asteraceae). A plant family is a broad grouping of related genera that share common characteristics. The Latin Name of a family usually ends in "aceae," e.g., Liliaceae (the lily family). Nomenclature is from Swink and Wilhelm, 1979. A phenological (flowering date) chart is included on pages v and vi so that the observer will know which plants are in bloom at any particular time.

Illustrations are intended to be the primary tool for identification in this book. Therefore, plant descriptions are brief and given with a minimum of technical terms. When technical terms are used they are defined in the glossary. Whenever two species are likely to be confused, more detailed differences are given using both plant description and additional illustration. Those who wish a more scientific approach to plant identification may consult a technical book, such as *Plants of the Chicago Region* by Swink and Wilhelm.

Ecological notes are presented for the three most prominent plant families, and for some genera and most species. These notes provide information about the general importance of the plants to the prairie community, plant pollination ecology, unique anatomical structures enabling the species to survive in the prairie community, and other points of interest.

PLANT NAMES

There is a human need to name plants. Once a person correctly learns the name and identification of a prairie plant, he/she is likely to remember some of its ecology and importance to the prairie community. Most plants have a common and a scientific name which are usually complementary and have definite meanings.

Common Names

The common name is often locally recognized and consists of everyday words; however, it is seldom universal. Unfortunately many plants have the same or similar common names, and in some cases a plant may have several common names. In this booklet, prairie plants are designated by the common names most frequently used in northern Illinois.

Scientific Names

Each plant has only one valid scientific name, i.e., the species. The scientific name is made up of three or more parts: (a) the genus, (b) the specific epithet, and (c) the authority who named the plant (Note: The author's name is often abbreviated). The genus (plural: genera) is a group of closely related species. The specific epithet is often referred to as the "species" but this is incorrect. The genus name and the specific epithet together form the name of the species. A species (the word "species" is both singular and plural) is a group of genetically and structurally related organisms which are capable of producing fertile offspring. An example of a scientific name is *Dodecatheon meadia* L.(the common name for shooting star). The genus is *Dodecatheon,* the specific epithet is *meadia*, and the author is L. (abbreviation for Linnaeus).

Sometimes a plant species has a recognizable subspecies or variety which is included. For example, *Heuchera richardsonii* var. *grayana* R. Br. is the scientific name for prairie alum root. The variety name for this plant is *grayana*. The genus, specific epithet, and variety should be underlined or italicized in written literature.

For good botanical/taxonomic reasons, plant names are occasionally changed to another specific epithet or genus following rules set forth by the International Code of Botanical Nomenclature. An example of a plant whose name has been changed is Culver's Root, *Veronicastrum virginicum* (L.) Farw. The name in parentheses is that of the person(s) who first applied the scientific name, followed by the person(s) who transferred the plant to another genus or gave it a different specific epithet.

It is sometimes not necessary or possible to give the specific epithet. In these cases the abbreviations, sp. (singular) and spp. (plural), are used. The abbreviations are not underlined or italicized. In this book the abbreviation "spp." is often used when referring to more than one species belonging to the same genus. For example the indigos are discussed as *Baptisia* spp.

PHENOLOGICAL (FLOWERING DATE) CHART

The solid line indicates the average blooming dates of the prairie plant species in this book. This chart does not include the earliest flowering or latest flowering dates during years having a very early spring or a very late autumn.

	APRIL	MAY	JUNE	JULY	AUG	SEPT	OCT
Allium cernuum, Nodding Wild Onion					——		
Amorpha canescens, Lead Plant				——			
Andropogon gerardi, Big Bluestem					————		
Andropogon scoparius, Little Bluestem					————		
Anemone cyclindrica, Thimbleweed			——				
Asclepias sullivantii, Prairie Milkweed			——				
Asclepias tuberosa, Butterfly Milkweed			——				
Aster ericoides, Heath Aster					——		
Aster laevis, Smooth Blue Aster					——		
Aster novae-angliae, New England Aster					————		
Baptisia leucantha, White Wild Indigo			——				
Baptisia leucophaea, Cream Wild Indigo		——					
Bouteloua curtipendula, Side-Oats Grama				——			
Bromus kalmii, Prairie Brome			——				
Cacalia tuberosa, Indian Plantain			——				
Calamagrostis canadensis, Blue Joint Grass			——				
Carex bicknellii, Prairie Sedge		——					
Ceanothus americanus, New Jersey Tea			——				
Coreopsis palmata, Prairie Coreopsis			——				
Coreopsis tripteris, Tall Coreopsis					————		
Cypripedium candidum, White Lady's Slipper		——					
Desmodium canadense, Showy Tick Trefoil				——			
Desmodium illinoense, Illinois Tick Trefoil				——			
Dodecatheon meadia, Shooting Star	——						
Echinacea pallida, Purple Coneflower			——				
Eryngium yuccifolium, Rattlesnake Master				——			
Euphorbia corollata, Flowering Spurge				——			
Gentiana andrewsii, Bottle Gentain					——		
Gentiana flavida, Yellowish Gentain					——		
Heuchera richardsonii grayana, Alum Root		——					
Hierochloe odorata, Vanilla Grass		——					
Hypoxis hirsuta, Yellow Star Grass		——					
Lespedeza capitata, Round-Headed Bush Clover				——			
Liatris aspera, Rough Blazing Star					——		

v

	APRIL	MAY	JUNE	JULY	AUG	SEPT	OCT
Liatris pycnostachya, Prairie Blazing Star					——		
Lithospermum canescens, Prairie Puccoon		——————					
Monarda fistulosa, Wild Bergamot			————————				
Panicum virgatum, Switch Grass				———————			
Parthenium integrifolium, Wild Quinine				——			
Petalostemum candidum, White Prairie Clover			——				
Petalostemum purpureum, Purple Prairie Clover			——				
Phlox glaberrima interior, Marsh Phlox			————				
Phlox pilosa, Prairie Phlox		——					
Potentilla arguta, Prairie Cinquefoil			——				
Psoralea tenuiflora, Scurfy Pea			——				
Pycnanthemum virginianum, Common Mountain Mint					————		
Ratibida pinnata, Yellow Coneflower				———————			
Rudbeckia hirta, Black-Eyed Susan			————————————				
Senecio pauperculus balsamitae, Balsam Ragwort		——					
Silphium integrifolium, Rosin Weed				———————————			
Silphium laciniatum, Compass Plant				———————			
Silphium perfoliatum, Cup Plant					————		
Silphium terebinthinaceum, Prairie Dock				————————————			
Sisyrinchium albidum, Blue-Eyed Grass		——					
Solidago nemoralis, Old-Field Goldenrod						———	
Solidago rigida, Rigid Goldenrod						———	
Sorghastrum nutans, Indian Grass					————		
Spartina pectinata, Prairie Cord Grass				——			
Sporobolus heterolepis, Prairie Dropseed					————		
Stipa spartea, Porcupine Grass		——————					
Tradescantia ohiensis, Spiderwort		————————					
Veronicastrum virginicum, Culver's Root				————			
Zizia aurea, Golden Alexander		——					

vi

APIACEAE (CARROT-PARSLEY)

Eryngium yuccifolium Michx., RATTLESNAKE MASTER

Height: 1-1.7 meters (3-5 feet)
Flowers: July-August
Color: White
Habitat: Mesic-Xeric

Identification Features:

The solitary stem of Rattlesnake Master grows from a short, thick rootstock. Thick, bayonet-shaped, alternate leaves clasp the stem. The leaves are parallel veined and have weak yucca-like bristles spaced far apart along the edges. The flower heads are white, hard, stiff and prickly. This elegant plant is easily recognized throughout the growing season.

Ecological Notes:

Rattlesnake Master is a member of the stable prairie community. It decreases in prairie where ungulate grazing occurs as its new growth is palatable and nutritious. The flowers attract many species of small insects for pollination activities. The leaves are rarely eaten by insects, such as grasshoppers and caterpillars, due to the strong cords of lignified cells running along the margins and veins. Anatomical features, such as waxy leaves and sunken pores in the leaf's upper epidermis, may help reduce the loss of water vapor from Rattlesnake Master during drought conditions.

Zizia aurea (L.) W. D. J. Koch, GOLDEN ALEXANDER

Height: 0.3-1 meter (1-3 feet)
Flowers: May-June
Color: Golden-Yellow
Habitat: Hydric-Mesic

Identification Features:

Golden Alexander has one to eight erect stems that are branched. Its basal and lower leaves are usually three-parted and have longer petioles than those near the top. The leaves are irregularly shaped, bright green, and have fine saw-toothed margins. The golden-yellow flowers are small and occur in round clusters known as umbels. The whole umbel is like a bouquet.

Ecological Notes:

Golden Alexander thrives in moist prairies and woods.

ASCLEPIADACEAE (MILKWEED)

Ecological notes and pollination mechanism for the Milkweeds (*Asclepias* spp.):

The flower of the milkweed is composed of five nectar horns, with keyhole-like niches between them, and five recurved sepals. Pollen sacs (pollinia), which are attached to a Y-shaped structure within the corona, are located in slits between the anthers. When an insect comes to gather nectar, it invariably slips its legs into a niche. The insect is held firmly while it sips nectar but as it struggles to free itself, attached pollinia collected from previously visited milkweeds are sloughed off and other pollinia are picked up. Usually the insect is able to free itself by pulling its foot upward and out. However, some insects are unable to free themselves and die on the flower.

The above described mechanism ensures cross-fertilization for the milkweed flowers. Although there are dozens of flowers in the cluster, usually only two or three become properly pollinated. The pollinated flowers form large downy pods full of silk-plumed seeds.

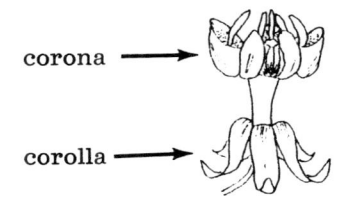

corona ⟶

corolla ⟶

MILKWEED FLOWER   POLLINIA

Asclepias sullivantii Engelm., PRAIRIE MILKWEED

Height: 0.7-1.5 meters (2-5 feet)
Flowers: Mid June-July
Color: Pink-Rose-Purple
Habitat: Hydric-Mesic

Identification Features:

The smooth, stout stem of Prairie Milkweed is pink to green-white in color. Its thick leaves are sessile (or nearly so), hairless, and point upward. They have conspicuous pink veins in addition to a strong, pink midrib. The umbels are terminal and wide-spreading. The individual blossoms are rose-pink in color and larger (1 cm) than those of the Common Milkweed.

Ecological Notes:

The presence of Prairie Milkweed often indicates virgin prairie conditions as it does not tolerate much disturbance. As with all milkweeds, its latex contains bitter alkaline chemicals which usually give the mature milkweed protection from most leaf and stem herbivores. There is a noticeable absence of insect holes in the stems and leaves of all the milkweeds. However, one of the few leaf and stem herbivores known to eat the acrid leaves of the Prairie and Common Milkweed is the larva of the Monarch Butterfly. The alkaline substance accumulates in the larva, and gives the adult butterfly some protection from bird predators.

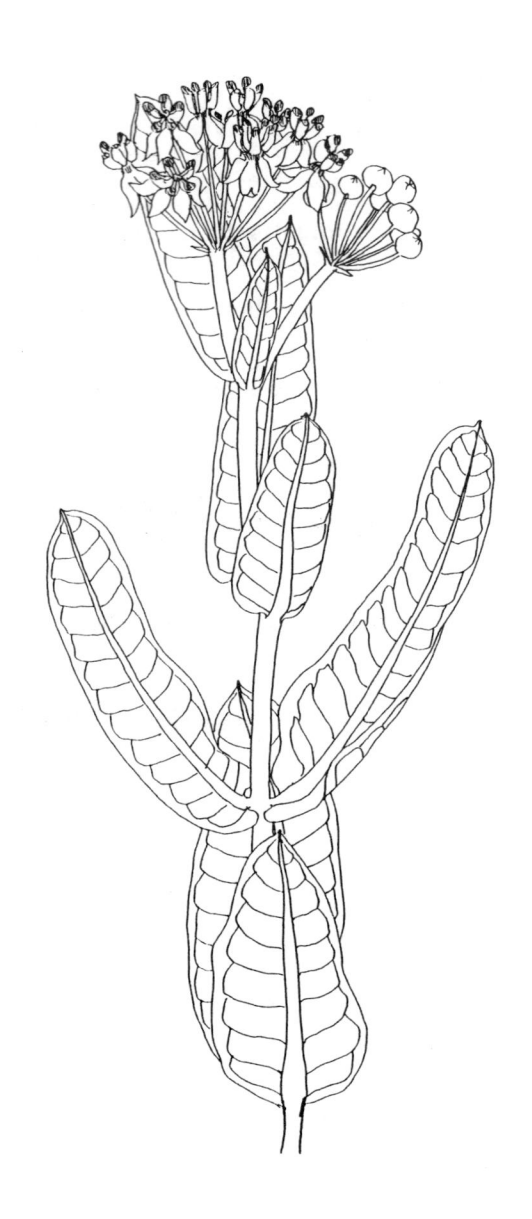

Asclepias tuberosa L., BUTTERFLY MILKWEED

Height: 0.3-0.8 meter (1-2.5 feet)
Flowers: Late June-Mid August
Color: Orange-Yellowish
Habitat: Xeric-Mesic

Identification Features:

Butterfly Milkweed has one to several erect hairy stems that arise from a stout tuberous taproot. It has lance-like to tapered alternate leaves that are strongly pubescent. Butterfly Milkweed has three to nine orange-yellowish flowers arranged in one to several umbels for each of its flat-topped inflorescences. Butterfly Milkweed differs from other milkweeds in that its stems and leaves do not excrete a white latex when cut. Butterfly Milkweed is one of the most spectacular prairie plants while in bloom and it attracts many insects.

Ecological Notes:

Butterfly Milkweed thrives along roadsides (especially where there is well-drained sandy soil), in savanna, and in prairie. Butterfly Milkweed is usually pollinated by flying insects as its hairy stems make it difficult for walking insects to climb up to the flowers. Ungulates do not eat Butterfly Milkweed; thus it tends to increase in degraded rangeland.

ASTERACEAE (COMPOSITE-SUNFLOWER)

Asteraceae (Composite-Sunflower Family)

General importance of the composite family to the prairie and other notes:

The composites have more species in the prairie flora than any other family. Members of this family contribute the most growth forms and beauty to the prairie community. Each stage of the growing season has a rich spectacle of composites in bloom.

The composites contribute to the soil building process. Some plants, such as Compass Plant, have thick taproots that penetrate deeply into the clay subsoil and help break it up. The rhizomes of composites, such as Asters, help bind the soil especially if it has been disturbed.

Many insects gather pollen from the showy composites. Some composites, such as the sunflowers, have large seeds that are eaten by grassland birds and small mammals. The thick roots of many composite members provide food for some small mammals during the late fall and winter months.

The composite head consists of two types of flowers: disk flowers and ray flowers. The smallest and most numerous flowers in the head are disk flowers which are generally located in the center. The strap-like "petals" surrounding the disk flowers are ray flowers. Many composites have both disk and ray flowers, while some have only one or the other.

Ecological notes for the asters, *Aster* spp.

One of the largest genera in the prairie community is *Aster*. Members of this genus occupy a great variety of habitats. For example, New England Aster and Heath Aster tend to be early colonizing species and also thrive in degraded prairie. Other asters, such as Smooth Blue Aster, are members of the stable prairie community.

Aster ericoides L., HEATH ASTER

Height: 0.3-0.6 meter (1-2 feet)
Flowers: September-Early October
Color: White rays, Yellow disks
Habitat: Mesic-Xeric

Identification Features:

Heath Aster usually grows in bush-like clumps from a root system which is highly rhizomatous. Its stem and leaves are covered with very short hairs. The numerous heads and leaves are tiny, especially toward the top of the plant, where they form a pyramid. The pyramid is 30-50 cm at the base. The composite head is about 5-6 mm in diameter and is composed of white ray flowers and yellowish disk flowers.

Ecological Notes:

Heath Aster is often abundant in prairies that are degraded due to factors such as overgrazing. Its abundance in degraded prairie is due to prolific seed production, low palatability, and extensive rhizome development. The dense rhizomatous root system helps bind the soil.

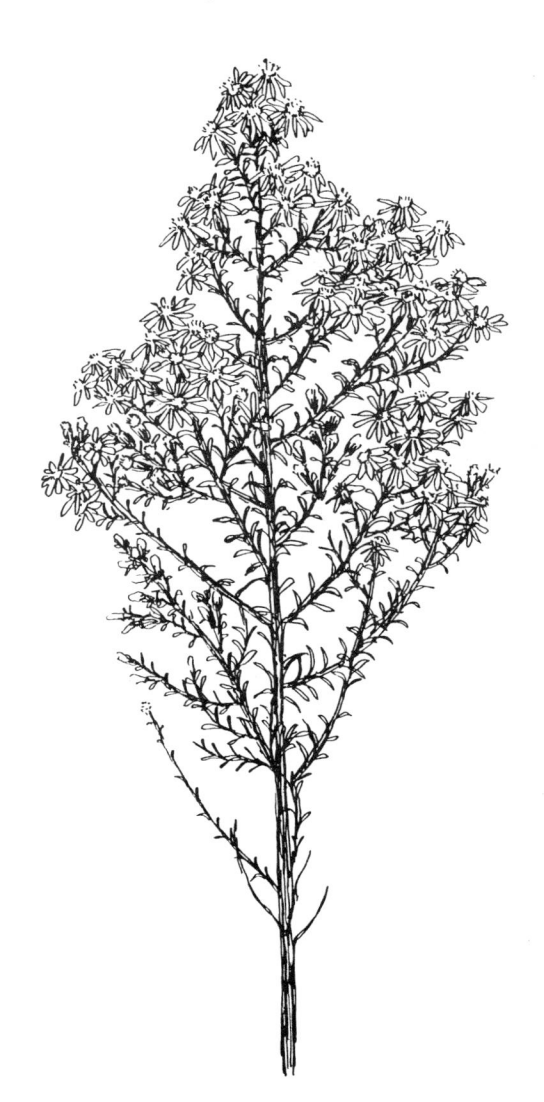

Aster laevis L., SMOOTH BLUE ASTER

Height: 0.3-1 meter (1-3 feet)
Flowers: Mid August-October
Color: Blue-Violet rays, Yellow disks
Habitat: Mesic-Xeric

Identification Features:

Smooth Blue Aster has an erect stem that arises from a short rhizome. The alternate leaves are smooth, blue-green and lanceolate. They have rough margins and clasp the stem. The basal leaves are much larger than those near the flowers. The flower heads are numerous with blue-violet ray flowers and yellow disk flowers.

Ecological Notes:

Smooth Blue Aster is a member of the stable prairie community.

Aster novae-angliae L., NEW ENGLAND ASTER

Height: 0.7-1.8 meters (2-5 feet)
Flowers: August-October
Color: Purple-Blue rays, Orange-Yellow disks
Habitat: Hydric-Mesic

Identification Features:

The tall, leafy stalks of New England Aster protrude in all directions from its rhizomes. The lanceolate-shaped leaves are quite uniform in size and strongly clasp the stem. The terminal inflorescences are branched and have several flower heads. These large aster heads are up to 2.5 cm in diameter with purple-blue ray flowers and reddish-orange-yellow disk flowers. The flowers of New England Aster add much color to the prairie during autumn.

Ecological Notes:

New England Aster is an early colonizing species and declines as the prairie community matures. It can also survive severe disturbances.

Cacalia tuberosa Nutt., INDIAN PLANTAIN

Height: 0.7-1.7 meters (2-5 feet)
Flowers: June-July
Color: Cream-White disks
Habitat: Mesic-Hydric

Identification Features:

Indian Plantain has an erect, smooth stem that arises from a small tuberous-thickened base and fleshy, fibrous roots. Its basal leaves are thick, oval, and have five to nine prominent parallel veins that converge toward the tip of the leaf blade. The basal leaves are often perforated with insect holes. The upper leaves are smaller and toothed near the tips. The flat-topped inflorescence is composed of white-cream flower heads with five tubular disk flowers. The achene has a very silky pappus.

Ecological Notes:

The presence of Indian Plantain indicates virgin prairie conditions. This species grows best in calcareous soils.

Coreopsis palmata Nutt., PRAIRIE COREOPSIS

Height: 0.6-0.9 meter (2-3 feet)
Flowers: Mid June-July
Color: Yellow-Orange rays and Brown disks
Habitat: Mesic-Xeric

Identification Features:

Prairie Coreopsis has a smooth, erect stem that arises from rhizomes. Its stiff leaves are opposite, sessile, palmately three-lobed (resembling a bird's foot) to the center, and about 5 cm long. The terminal inflorescence flower heads are up to 5 cm in diameter and contain both yellow ray and disk flowers. The ray flowers are often three-toothed at the tip. The foliage of Prairie Coreopsis turns to a beautiful orange-purple color during autumn.

Ecological Notes:

Prairie Coreopsis is highly rhizomatous and forms colonies that often exclude other species. Flowering usually occurs only on the perimeter of these colonies. The presence of Prairie Coreopsis often indicates virgin prairie conditions.

Coreopsis tripteris L., TALL COREOPSIS

Height: 1-2 meters (3-7 feet)
Flowers: August-September
Color: Yellow rays and Brown disks
Habitat: Mesic-Hydric

Identification Features:

The tall, smooth stem branches at the top and is capped by numerous flower clusters. The principle leaves are usually divided to the midrib into three leaflets. The yellow ray flowers are very showy. During fall, the leaves and stem of Tall Coreopsis turn a beautiful reddish-orange color.

Ecological Notes:

Tall Coreopsis grows under many conditions and is almost weedy. The flowers produce much pollen for bees and other pollinators at a time when there are not many forbs blooming in the prairie.

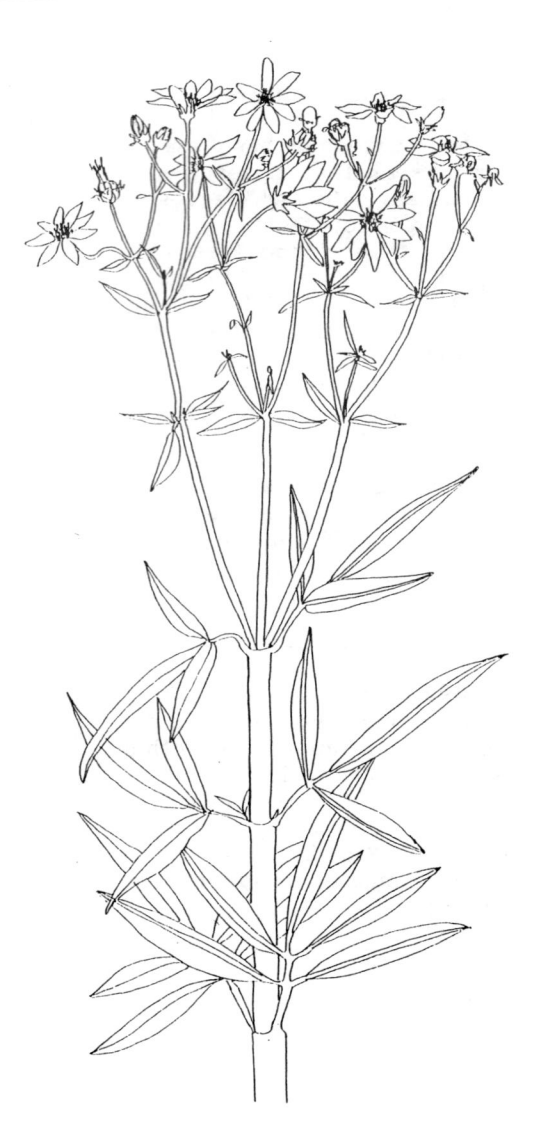

Echinacea pallida Nutt., PURPLE CONEFLOWER

Height: 0.5-0.9 meter (1.5-3 feet)
Flowers: June
Color: Purple rays, Brownish-Black disks
Habitat: Mesic

Identification Features:

Purple Coneflower usually has two to five unbranched pubescent stems that have a single terminal flower head. It has a deep, thick (1-3 cm in diameter), tuberous taproot. Its basal leaves are quite large (up to 18 cm long and 3-4 cm wide), ovate-lanceolate in shape, noticeably pubescent, and strongly three-veined. Smaller leaves ascend the stem. Large pinkish-purple flower heads extend up to 10 cm in diameter when first in bloom. However after approximately one week, the pinkish-purple ray flowers begin to droop and form an inverted "cone." After approximately three weeks, the ray flowers drop off and the central cones of spiny disk flowers remain conspicuous until late autumn. This elegant plant is easily recognized throughout the growing season.

Ecological Notes:

Purple Coneflower is an indicator of high quality virgin prairie. This colorful forb is palatable and nutritious to ungulates. The roots are a preferred food source of many rodents such as voles. While it is in bloom, many insects seek out the yellow pollen from the disk flowers.

Ecological notes for the blazing stars, *Liatris* spp.

Blazing stars tend to be early colonizing plants that can thrive in degraded or recovering prairie communities. When compared to other early colonizers, they have a relatively long vegetative life. Blazing stars are well adapted to periods of drought. They have a deep root system for water absorption, small hairs that help reflect sunlight, and numerous narrow leaves which help cool the plant, thus reducing transpiration.

Inflorescences that occur on spikes for most plant species start to bloom at the base and progress upward; but blazing stars bloom from the top and proceed downward. While in bloom, the "cattail" of rose-purple disk flowers is fuzzy with extended white stamens and pistils.

Most of the pollination observed for the blazing stars is carried out by a bumblebee, *Bombus pennsylvanicus*, a few butterfly species, beetles, flies, and carpenter bees. Bees rarely revisit flower heads while working an inflorescence, thus extracting maximum energy from the limited food resources and ensuring cross-pollination.

Various herbivores, in addition to insects, obtain energy from the blazing stars. Ungulates graze the young plants. Rodents, such as the Meadow Vole, seek out and eat the corms.

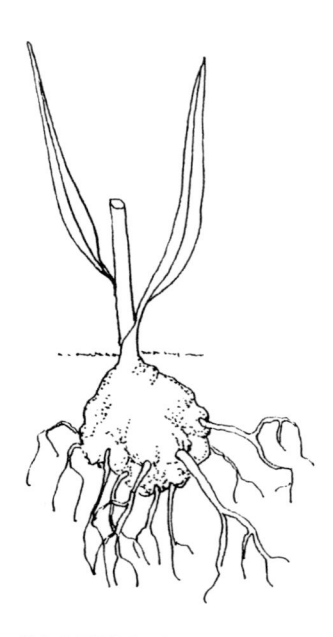

BLAZING STAR CORM

Liatris aspera Michx., ROUGH (BUTTON) BLAZING STAR

Height: 0.3-0.6 meter (1-2 feet)
Flowers: Mid August-September
Color: Rose-Purple disks
Habitat: Xeric-Mesic

Identification Features:

Rough Blazing Star has an erect stem that develops from a large corm. The alternate leaves are numerous, rough, and narrow along the entire length of the stem. They diminish in size toward the summit. The leaves and stems are pubescent. The involucral bracts are curved in at the edges and appear cup-shaped or "button-like." The flower heads of Rough Blazing Star are larger than those of Prairie Blazing Star but not as numerous.

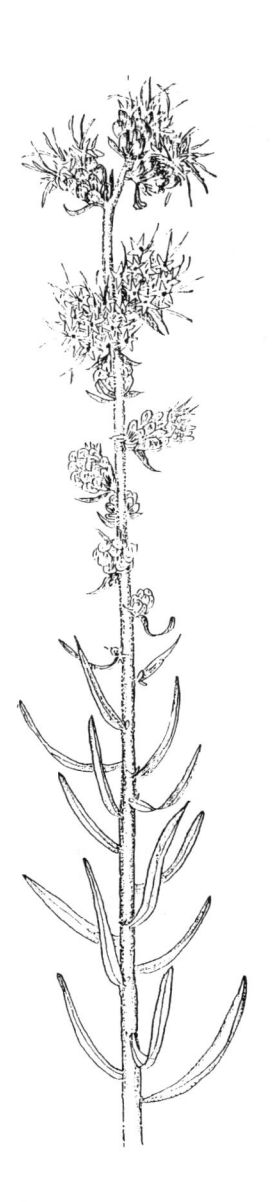

Liatris pycnostachya Michx., PRAIRIE BLAZING STAR

Height: 1-1.5 meters (3-5 feet)
Flowers: Mid July-August
Color: Rose-Purple disks
Habitat: Mesic-Hydric

Identification Features:

The stem of Prairie Blazing Star is closely set with spirals of long, narrow leaves near the base that progressively get shorter toward the inflorescence spikelet. Its unbranched stem arises from a large corm that is 7-10 cm in diameter. Short, stiff hairs occur on both the leaves and stem. The involucral bracts are reddish, sharp-pointed, fuzzy, and bent out. The cylindrical inflorescence is a spikelet, 25-30 cm long, and crowded with sessile heads of disk flowers.

Parthenium integrifolium L., WILD QUININE

Height: 0.3-1 meter (1-3 feet)
Flowers: Mid June-August
Color: White rays and disks
Habitat: Mesic

Identification Features:

The stems of Wild Quinine arise from a thick tuberous taproot. Its leaves are rough, firm, serrated, and ovate-oblong in shape. The basal and lower leaves are large (30 cm long and 10 cm wide) with long petioles, while the upper ones are much smaller and clasp the stem. The beautiful white flower heads are numerous and in dense terminal clusters. Each flower head is about 4-5 mm in diameter with five tiny ray flowers and many disk flowers.

Ecological Notes:

Wild Quinine is a member of the stable prairie community and is rarely found in disturbed sites. Many small insects visit the white flower heads. The thick roots are a favorite food of the Meadow Vole.

Ratibida pinnata (Vent.) Barnh., YELLOW (GRAY-HEADED) CONEFLOWER

Height: 0.7-1.3 meters (2-4 feet)
Flowers: July-August
Color: Yellow rays, Greenish-Gray disks
Habitat: Mesic-Xeric

Identification Features:

The stems of Yellow Coneflower are usually clustered, branched, and have deeply dissected leaves with three to seven segments. The stem arises from a shallow rhizome. The basal leaves have short petioles while the upper leaves are nearly sessile. All the leaves are smooth. The terminal inflorescence consists of four to ten drooping yellow ray flowers and a central head of greenish-gray disk flowers. Identification of this unique flower is easy due to its rather ragged appearance, gray-green deeply dissected basal leaves, drooping yellow ray flowers, and a central head of greenish-gray disk flowers.

Ecological Notes:

Yellow Coneflower is a short-lived, colonizing plant (almost weedy) of the prairie community. It has a shallow rhizomatous root system that helps bind soil. Yellow Coneflower, when young, is highly palatable for ungulates.

Rudbeckia hirta L., BLACK-EYED SUSAN

Height: 0.3-0.6 meter (1-2 feet)
Flowers: Mid June-September
Color: Orange-Yellow rays, Brown disks
Habitat: Mesic

Identification Features:

Black-eyed Susan has highly branched, bristly-pubescent stems. Its alternate leaves are oblong, shallow-toothed, and pubescent. Black-eyed Susan has orange-yellow ray flowers and a central head of brown-black disk flowers studded with yellow pollen. Its inflorescences are up to 7 cm in diameter. Black-eyed Susan is very showy and remains in flower throughout the summer.

Ecological Notes:

Black-eyed Susan is a biennial or short-lived perennial that is an early colonizer of the prairie community. It often thrives in disturbed areas such as dusty roadsides. The leaves have long, stout hairs that serve to keep dust from clogging their breathing pores (stomata), thus helping to keep the plant alive and functioning properly.

Senecio pauperculus balsamitae (Muhl.) Fern., BALSAM RAGWORT

Height: 0.1-0.5 meter (4-18 inches)
Flowers: Mid May-June
Color: Yellow-Orange rays and disks
Habitat: Xeric-Mesic

Identification Features:

Balsam Ragwort is a small erect plant that arises from a branching rhizome. Its basal leaves are slender to oval with long petioles. The upper leaves are sessile and have deeply jagged and serrated lobes. There are few leaves on the stem. The flat-topped inflorescences at the tip of the stem are up to 2 cm in diameter. Both the ray and disk flowers are yellowish-orange. Balsam Ragwort adds much brightness to the late spring prairie scene.

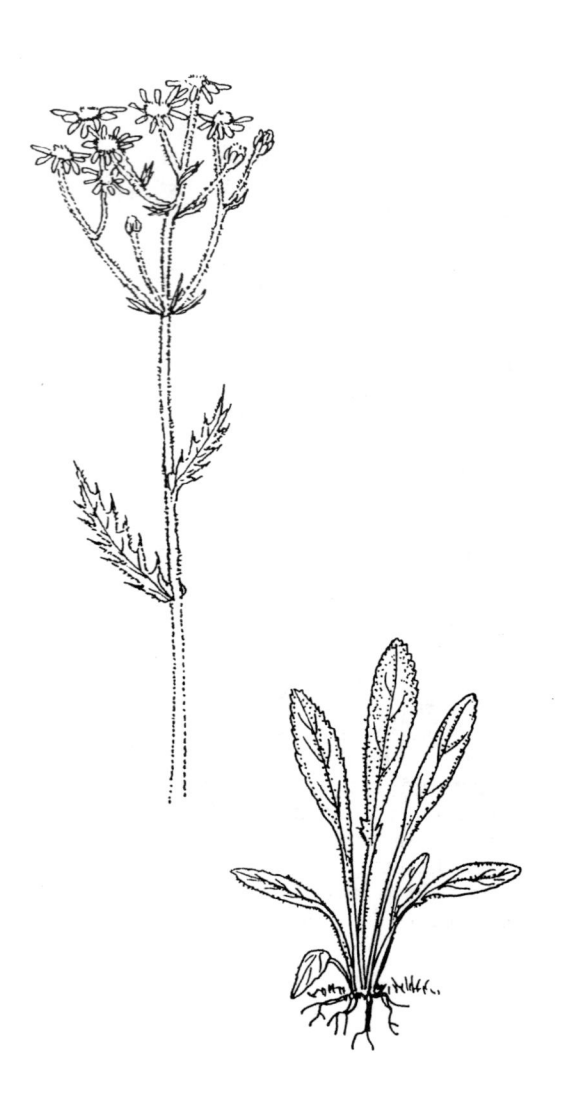

Ecological Notes:

Balsam Ragwort can survive in disturbed areas and is also a member of the stable prairie.

Ecological notes for the *Silphium* species:

Silphiums are large, coarse forbs characteristic of the prairie community. The large root systems of silphiums are important in the soil building process as they can penetrate clay subsoil. The thick taproots (2.5-5 cm in diameter) of Compass Plant and Prairie Dock descend vertically, start branching at 1 m down, and reach a depth of 3-4.5 m. The root systems of Rosinweed and Cup Plant are more fibrous and dense. As space between the soil and root increases due to root contraction during flowering or soil contraction during periods of drought, humus seeps into these crevices and black soil forms.

All silphiums are palatable and nutritious during early growth and are sought by ungulates. Silphiums look like sunflowers. They have both ray and disk flowers, all yellow. They produce large sunflower-like seeds that are actively sought by birds and rodents.

Moisture does not easily evaporate from the large silphium leaves due to their rough, waxy layer and orientation. In particular, the leaves of Compass Plant and Prairie Dock are erect above the ground and tend to present only their thin edges to the sun, thus reducing transpiration. The large leaves of these two plants feel cool even on hot days.

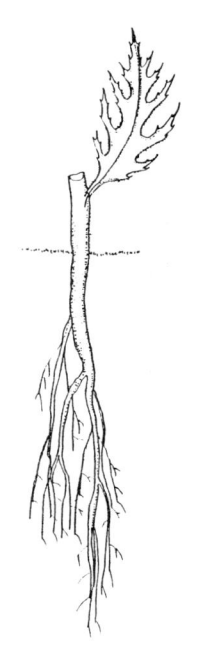

COMPASS PLANT
SEEDS

COMPASS PLANT
ROOT

Silphium integrifolium Michx., ROSINWEED

Height: 0.7-1.5 meters (2.5-5 feet)
Flowers: July-August
Color: Yellow rays and disks
Habitat: Mesic

Identification Features:

The stem of Rosinweed arises from rhizomes. Its leaves, as with other members of this genus, feel like sandpaper. They are opposite, lance-olate to ovate, and slightly toothed. The yellow terminal heads (5 cm in diameter) are composed of both ray and disk flowers.

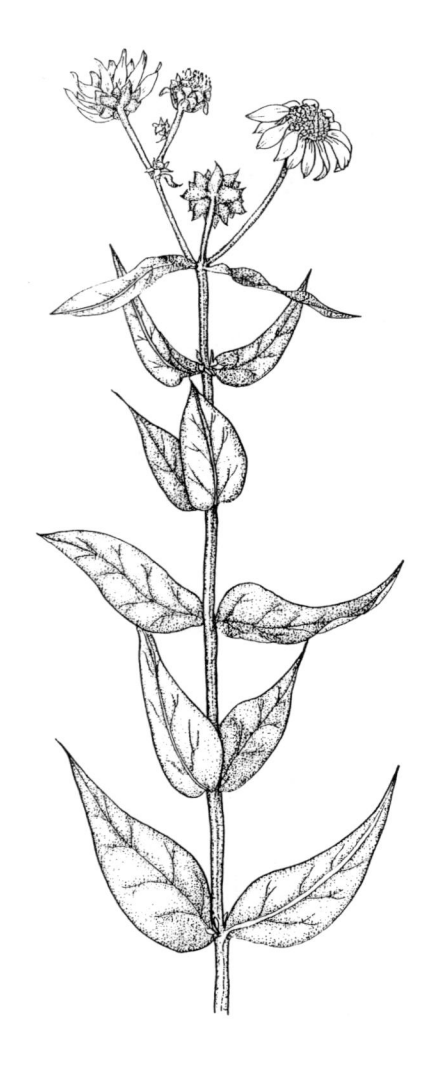

Silphium laciniatum L., COMPASS PLANT

Height: 1-2.5 meters (3.5-8 feet)
Flowers: July-August
Color: Yellow rays and disks
Habitat: Mesic-Xeric

Identification Features:

The stem of Compass Plant is stout and bristly-hairy. The leaf edges, especially for plants that are not going to produce flowers, tend to point north and south, hence the common name. The basal leaves are large, deeply cut, and rough to the touch.
The upper leaves of the stem are reduced. The plant is resinous throughout with globules of white resin often oozing out and sticking to the stem. The upper half of the stem contains large (5-7 cm in diameter) bright yellow "sunflowers" having both ray and disk flowers.

Ecological Notes:

In the genus *Silphium*, Compass Plant has the greatest fidelity to undisturbed prairie communities. It is actively sought by native and domestic ungulates. A host specific weevil, *Rhynchites* sp., chews the stem near the flower heads so that they droop and wilt. Next it starts eating the disk flowers.
Resin then oozes out of the stem, and the flower heads die.

Silphium perfoliatum L., CUP PLANT

Height: 1.2-2.1 meters (4-7 feet)
Flowers: Late July-August
Color: Yellow rays and disks
Habitat: Hydric

Identification Features:

The bases of the large opposite leaves are grown together and surround the square stem to form a "cup." The large toothed leaves are triangular-shaped. Cup Plant has numerous heads with both ray and disk flowers, all yellow.

Ecological Notes:

In addition to growing in prairies, Cup Plant also thrives in flood plains and degraded areas. Following a rain, considerable amounts of water are held in the "cups." Birds have been observed drinking water from the "cups" of the Cup Plant.

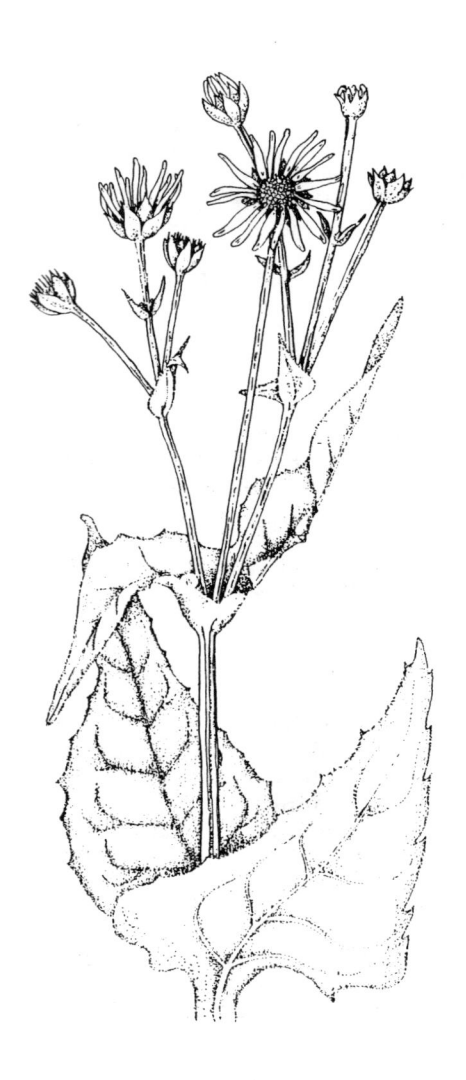

Silphium terebinthinaceum Jacq., PRAIRIE DOCK

Height: 0.7-3 meters (2-10 feet)
Flowers: July-September
Color: Yellow rays and disks
Habitat: Mesic-Hydric

Identification Features:

The leaves are 30 cm or longer, thick, resinous, coarsely toothed, spade-shaped, and rough like sandpaper. They are confined to the base of the plant and, like other silphiums, are coated with a rough, waxy layer. Usually during August, tall, smooth stalks arise from the cluster of leaves. At the top of these stalks there are about six smooth, round buds that open as small "sunflowers." The resin in the leaves and stems has an odor suggestive of turpentine, hence the Latin name *terebinthinaceum*.

Ecological Notes:

New shoots from the large taproot help Prairie Dock to survive severe degradation. It is sometimes the only remaining prairie plant species surviving in an area.

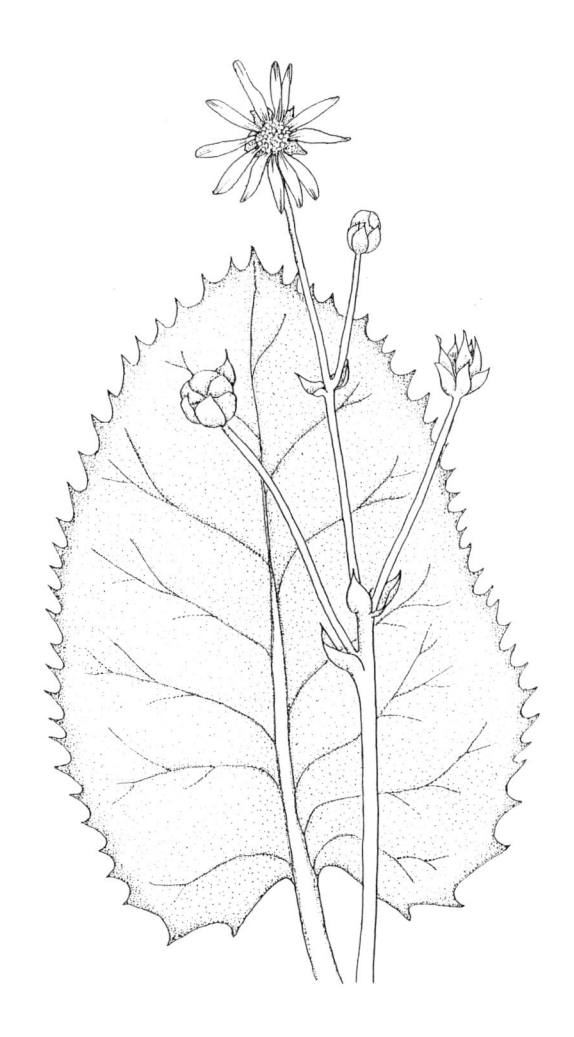

Ecological notes for the goldenrods, *Solidago* spp.

Goldenrods are common and hardy in prairie and in disturbed communities. The palatability of most goldenrods is low; thus they often increase in overgrazed prairies. Goldenrods have a dense rhizomatous root system that helps binds the soil.

The pollen is sticky and spread by insects, not wind. It is *not* a source of hay fever as commonly accused. Certain species of soldier beetles, resembling brightly colored fireflies, feed on the goldenrod heads and carry their pollen from plant to plant.

Solidago nemoralis Ait., OLD-FIELD GOLDENROD

Height: 0.2-0.6 meters (0.5-2 feet)
Flowers: Mid August-September
Color: Yellow rays and disks
Habitat: Xeric

Identification Features:

Old Field Goldenrod has upright, unbranched smooth stems that arise from its rhizomes. The firm, narrow leaves are three-veined, alternate, and clasp the lower portion of the stem. They are up to 15 cm long at the base and are much smaller upward toward the flower heads. The inflorescence is a recurving one-sided raceme with the heads pointing outward. The small heads have yellow ray and disk flowers.

Ecological Notes:

The dense rhizomatous root system of Old Field Goldenrod helps bind the soil. The narrow leaves of this species help radiate heat more readily, thus reducing some transpiration.

Solidago rigida L., RIGID (STIFF) GOLDENROD

Height: 0.3-1.7 meters (1-5 feet)
Flowers: Late August-September
Color: Yellow rays and disks
Habitat: Xeric-Mesic

Identification Features:

Rigid Goldenrod has a rigid, hairy stem that arises from a short rhizome. Its stiff stem branches only at the top and is capped by dome-shaped to flat-topped clusters of small heads of yellow-orange ray and disk flowers. The inflorescence is about 15 cm in diameter. The alternate leaves are thick and leathery, rigid, pubescent on both surfaces, and have pointed tips. The lower leaves are large (up to 30 cm long) and have short petioles, while the upper ones are smaller and clasp the stem.

Ecological Notes:

Rigid Goldenrod is a colonizing species that often invades disturbed areas. Rigid Goldenrod competes well with grasses for moisture as its fibrous roots can penetrate the soil to depths of 5 m. Nitrogen fixation has been reported for Rigid Goldenrod but at a much lower rate than for legumes. It is unclear as to whether nitrogen fixation in this species is a biologically important source of nitrogen for individual plants.

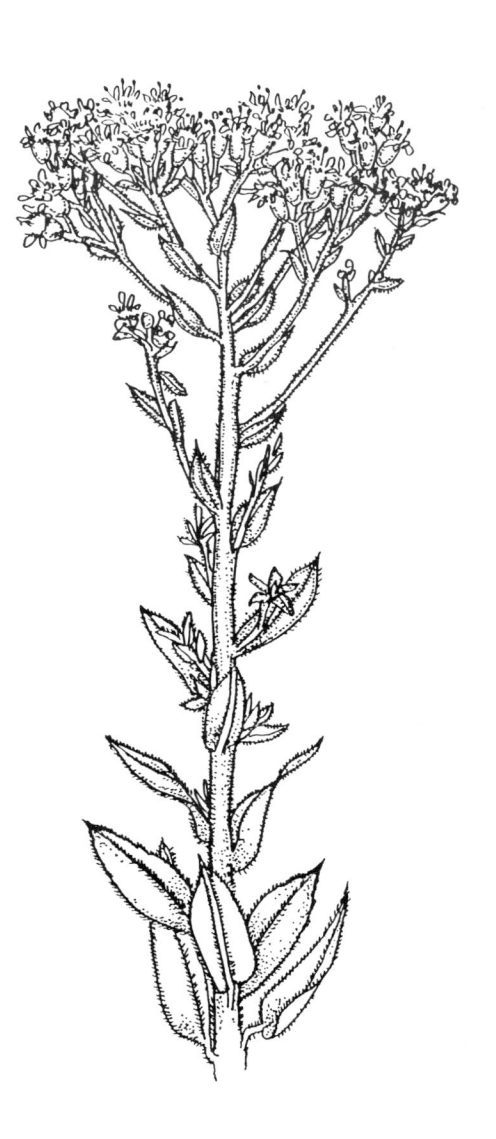

BORAGINACEAE (BORAGE)

Lithospermum canescens (Michx.) Lehm., PRAIRIE PUCCOON

Height: 0.2-0.3 meter (8-12 inches)
Flowers: May-June
Color: Yellow-Orange
Habitat: Xeric-Mesic

Identification Features:

One to several pubescent stems grow from a long taproot that contains reddish-purple juices. The alternate leaves are lance-shaped to oblong and are covered with short, stiff hairs. The stems are topped with inflorescences of tubular yellow-orange flowers that have five lobes. Prairie Puccoon glitters in sunlight! Its seeds are like little pieces of polished bone or ivory.

Ecological Notes:

The presence of Prairie Puccoon is an indication that virgin prairie conditions exist. The tube of the flower is constructed in such a manner that only a few kinds of insects can enter to gather nectar.

COMMELINACEAE (SPIDERWORT)

Tradescantia ohiensis Raf., SPIDERWORT

Height: 0.3-1 meter (0.9-3 feet)
Flowers: Late May-July
Color: Bluish-Violet
Habitat: Mesic-Hydric

Identification Features:

The stems of Spiderwort arise from numerous fleshy roots. Its alternate leaves are long (up to 45 cm) and narrow. They are folded lengthwise and curve downward. The stems contain a mucilaginous, stringy substance resembling that excreted by a spider. The flowers have three round to spade-shaped petals accented by six golden stamens.

Ecological Notes:

This plant needs sunshine, but not too much. During intense sunlight the silky petals of the flower close. The mucilaginous, stringy substance enables Spiderwort to hold considerable amounts of water which may help it survive periods of drought.

CYPERACEAE (SEDGE)

Carex bicknellii **Britt.**, PRAIRIE (BICKNELL'S) SEDGE

Height: 0.3-0.8 meter (1-2.5 feet)
Flowers: May
Color: Yellowish anthers
Habitat: Xeric-Mesic

Identification Features:

The erect stems of Prairie Sedge usually grow in narrow, upright tufts. Its stem has three-angles. The stems of sedges are not round like those of grasses. The perigynium (an inflated sac that encloses the achene) is 4 mm or more in length.

Ecological Notes:

Prairie Sedge is a stable member of xeric and mesic prairie.

EUPHORBIACEAE (SPURGE)

Euphorbia corollata L., FLOWERING SPURGE

Height: 0.3-1.2 meters (1-4 feet)
Flowers: July-August
Color: White
Habitat: Xeric-Mesic

Identification Features:

The stem of Flowering Spurge is slender and smooth. Its leaves are oblong, 1-2 cm long, and scattered alternately along the stem. The base of each inflorescence has several leaves in a whorl. The flowers have petal-like appendages but no true petals. The fruit is a broad capsule which breaks open with enough force to send the seed a considerable distance. The entire plant contains milky sap.

Ecological Notes:

Flowering Spurge can become slightly weedy in some disturbed prairie habitats. It is a poisonous plant which is seldom eaten by ungulates. Flowering Spurge has thick, milky sap which holds considerable amounts of water. This sap, along with its deep root system, helps it survive drought periods.

FABACEAE (LEGUME)

General importance of the legume family to the prairie and other notes:

The legumes are a major family in the prairie flora. Legume species aid in the soil building process by means of their deep penetrating root system and the nitrogen-fixing bacteria (Genus *Rhizobium*) which live in their root nodules. *Rhizobium* species convert inert nitrogen from the soil's atmosphere into compounds that are usable by the legumes and by adjacent members of other plant species. These nitrogen-fixing bacteria are genus-specific for the plant; e.g., in the root nodules of lead plant, *Amorpha canescens*, live *Rhizobium amorpha*. The deep root systems of legumes aid in breaking up clay subsoil.

Some members of the legume family, such as the indigos, are very showy. Legumes produce hard seeds that remain viable in the soil for many years.

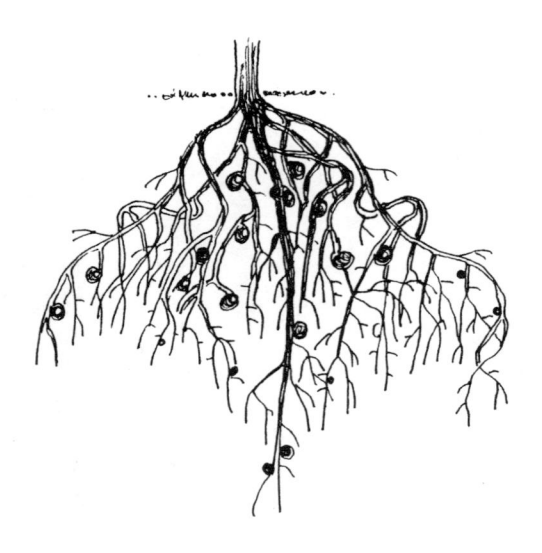

LEGUME ROOT SYSTEM
WITH NODULES

Amorpha canescens Pursh, LEAD PLANT

Height: 0.6-1.3 meters (2-4 feet)
Flowers: Late June-July
Color: Blue-Violet-Purple
Habitat: Mesic-Xeric

Identification Features:

Lead Plant is a true prairie shrub. Its stems are woody and slightly pubescent. The alternate leaves have 15-51 small leaflets (6-12 mm long), are gray-green (lead) colored, and are covered with fine hairs. The inflorescence is a narrow, crowded, highly branched raceme 5-10 cm long. Its flowers show only a single petal and have ten bright yellowish-orange conspicuous anthers. Lead Plant provides striking beauty to the prairie throughout the growing season.

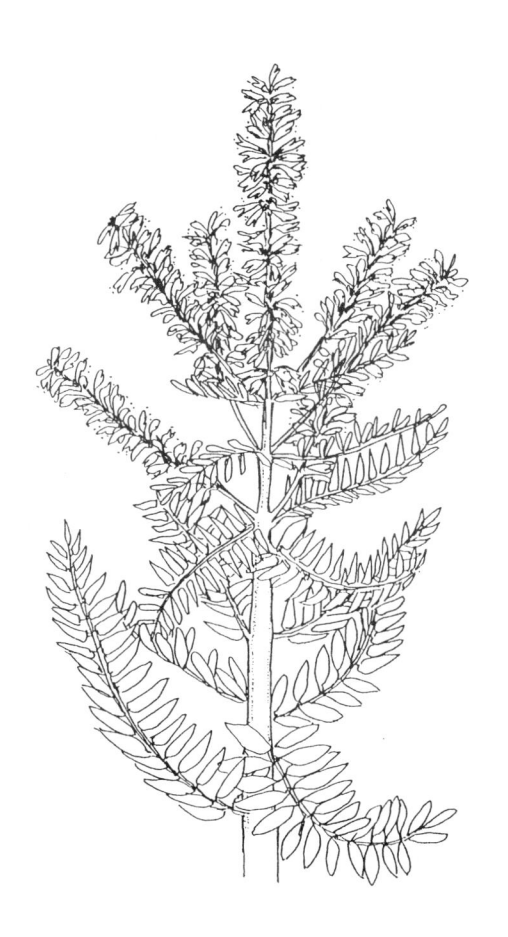

Ecological Notes:

Lead Plant is an indicator of virgin prairie. It is highly nutritious and palatable, and thus decreases under heavy grazing pressure. In the absence of fire or heavy grazing the stems become very woody. Lead Plant has several anatomical features which help it survive periods of drought. These include: (1) a deep root system for water absorption, (2) lead-colored leaves which reduce the effect of solar heating by reflecting sunlight, and (3) finely divided leaflets which expose less surface area to sunlight; thus transpiration is reduced.

Ecological notes for the indigos, *Baptisia* spp.

The indigos, *Baptisia* spp., are important soil builders. They have long penetrating roots which aid in breaking up clay subsoil. Living in their root nodules are the nitrogen-fixing bacteria, *Rhizobium baptisia*, which convert inert nitrogen into usable compounds. Although nutritious, the vegetative plant parts are rarely eaten due to their slight toxicity. Most of their large hard seeds are eaten or destroyed by insects. White Wild Indigo grows in prairie and savanna, whereas Cream Wild Indigo grows primarily in open prairie.

Cream Wild Indigo and White Wild Indigo are remarkable from both the standpoint of beauty and their insect-plant relationships. Described below are some pollination and seed predation aspects for these two prairie legumes.

Cream Wild Indigo blooms approximately two weeks earlier than White Wild Indigo. It is pollinated by bumblebee queens, *Bombus bimaculatus* and *B. nevadiensis auricomis*, that have just emerged from overwintering. The queens are also engaged in nest building, egg-laying, and rearing of immatures in addition to gathering nectar during this time. White Wild Indigo blooms approximately two weeks later and is generally pollinated by the worker bumblebee caste. Both species have a few visits from the honeybee, *Apis* sp. Ruby-throated Hummingbirds have also been observed gathering nectar from White Wild Indigo, but they are not a major pollinator.

The foraging (nectar gathering) pattern of bumblebee pollinators determines the likelihood of pollen transfer within the raceme. Indigos often have flowers in both the pistillate and staminate phase. Queens walk horizontally between the flowers of Cream Wild Indigo's raceme, moving from staminate phase flowers to pistillate phase flowers, and by so doing release pollen. Pollinator movement for White Wild Indigo is upward and reverse, e.g., the workers move from the pistillate to the staminate phase flowers.

White Wild Indigo produces a somewhat greater nectar reward than Cream Wild Indigo. Nectar production for White Wild Indigo is greater on days 3-4 of flowering than on day 1; there is no significant difference for Cream Wild Indigo. Frequency of pollination is usually higher in White Wild Indigo for three major reasons: (1) more bumblebees are present, (2) nectar rewards are higher, and (3) it is less subject to unpredictable weather conditions since flowering occurs later. As a result, more ovules (eggs) are fertilized and eventual seed predation is higher.

Several insects prey upon the seeds and pods of the indigos. Predation includes seed destruction, pod exit holes, webby material, and wall damage. In some years, there is almost a total loss of seeds except for a few isolated plants. *Apion rostrum* (a weevil) is the dominant seed predator. In early June, the weevils feed on the flower and leaves.

The female drills holes into the base of the inflated pod throughout the flowering season, lays her eggs, and then pushes the eggs inside the pod with her snout. The egg is yellow and only slightly smaller than the seeds at this time. Individual weevils inhabit the pods for varying time periods. Exit holes (1.5 mm in diameter) appear throughout the growing season; some adults leave when the pods dehisce (split), and others overwinter.

In some cases, the weevils may actually hasten seed germination by slightly gnawing the tough seed coat. Indigos disperse their seeds in a tumbleweed manner during late fall when the abscission layer at the stem base breaks. The stem, with its pods still attached, can then be blown by the wind.

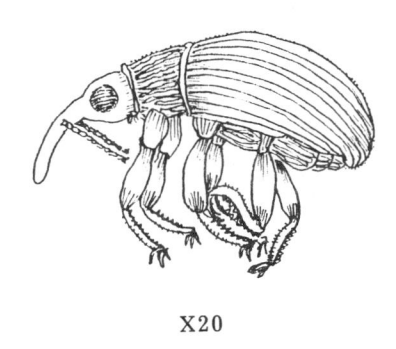

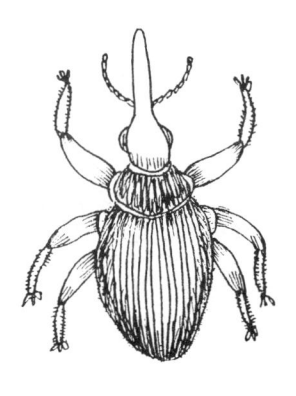

X20

X20

Apion rostrum

Baptisia leucantha T. & G., WHITE WILD INDIGO

Height: 1-1.7 meters (3-5 feet)
Flowers: June-July
Color: White
Habitat: Mesic-Xeric

Identification Features:

White Wild Indigo has a smooth, stout stem with a widely branched crown. It resembles a miniature bushy tree. Its smooth leaves have small stipules, short petioles, and three leaflets (1-2 cm long). The whole plant turns black upon drying. A tall, elegant inflorescence raceme (30-60 cm) stands erect above the bushy plant. The inflorescence is closely set with white pea flowers about 2.5 cm long. The smooth, inflated legume pod is 2-3 cm long and has a short (up to 0.5 cm) tapering beak.

FLOWER

Baptisia leucophaea Nutt., CREAM WILD INDIGO

Height: 0.5-0.9 meter (1.5-3 feet)
Flowers: May-Early June
Color: Cream-Yellow
Habitat: Mesic-Xeric

Identification Features:

This low, drooping bushy plant has two to twelve finely pubescent thick stems which are highly branched and widely spread (up to 0.8 m). The leaves are sessile, finely pubescent, and have two large stipules which give the appearance of leaflets. Cream Wild Indigo has only three leaflets (up to 4 cm long), but there appear to be five due to the two large stipules. The inflorescence consists of a horizontal to drooping raceme of large cream-yellow pea flowers that are up to 3 cm long. The finely pubescent inflated pod is 4-5 cm long and has a long, tapering beak (up to 1.5 cm long). Cream Wild Indigo is a most handsome plant at all stages during the growing season.

Ecological notes for the tick trefoils, *Desmodium* spp.

The Tick Trefoils, *Desmodium* spp., have one of the most efficient seed dispersal mechanisms of any prairie species. The "hooked" hairs on the loment pod cling to any animal or person passing by and are thus dispersed. The seeds provide nourishment for grassland birds and small mammals. As on most legumes, nitrogen-fixing bacteria live in root nodules and convert inert nitrogen into usable compounds for the legume and adjacent members of other plant species. The deep taproots aid in breaking up clay subsoil. Illinois Tick Trefoil has a greater fidelity to the prairie than Canada Tick Trefoil which is also found in disturbed habitats.

Desmodium canadense (L.) DC., CANADA (SHOWY) TICK TREFOIL

Height: 0.6-1.8 meters (2-5 feet)
Flowers: July-August
Color: Purple
Habitat: Mesic-Hydric

Identification Features:

Canada Tick Trefoil has a more bushy appearance than Illinois Tick Trefoil. The stems and leaflets are highly pubescent and have hooked hairs. Its leaf petioles are shorter than those of Illinois Tick Trefoil. The inflorescences are highly branched racemes of purple flowers that are on short pedicels. The flowers are numerous and dense.

Desmodium illinoense Gray, ILLINOIS TICK TREFOIL

Height: 0.6-1.7 meters (2-5 feet)
Flowers: July-August
Color: Pink-Light Purple
Habitat: Xeric-Mesic

Identification Features:

The tall erect stem of Illinois Tick Trefoil is finely pubescent and has hooked hairs. Its alternate leaves have a long petiole and three leaflets. They are pubescent with hooked hairs beneath. The inflorescence is a tall, sparsely branched raceme with pink to light purple pea flowers. The legume pod is a loment that breaks into one-seeded sections with hooked hairs.

LOMENT

Lespedeza capitata Michx., ROUND-HEADED BUSH CLOVER

Height: 0.6-1.5 meters (2-5 feet)
Flowers: Mid August-September
Color: White
Habitat: Mesic-Xeric

Identification Features:

Round-Headed Bush Clover has an erect, silvery-pubescent stem. Its alternate leaves have three leaflets which are oblong to lance-like in shape. They are relatively smooth on the upper side but silvery-pubescent beneath. The inflorescences are clustered heads (up to 3 cm in diameter) of small cream to white pea-shaped flowers near the top of the stems. The seed heads turn brown in the fall.

Ecological Notes:

Round-Headed Bush Clover grows in prairie and savanna. It is excellent forage for ungulates and decreases with heavy grazing. The seeds provide food for grassland birds.

Ecological notes for the prairie clovers, *Petalostemum* spp.

In northern Illinois, *Petalostemum* species are indicators of virgin prairie. They are highly sought after by herbivores, such as rabbits and ungulates, and decrease with grazing pressure. The prairie clovers have anatomical features that help them survive in times of water stress. Their finely divided leaflets offer less surface area to the sun; thus the plant is cooler and transpiration is reduced. A wide-ranging root system from the taproot helps absorb water. Nitrogen-fixing bacteria (*Rhizobium petalostemum*) in the root nodules convert inert nitrogen into usable compounds for the clovers and neighboring members of other plant species.

Petalostemum candidum (Willd.) Michx., WHITE PRAIRIE CLOVER

Height: 0.3-0.6 meter (1-2 feet)
Flowers: Late June-July
Color: White
Habitat: Xeric-Mesic

Identification Features:

White Prairie Clover has few to several smooth, upright stems arising from a taproot with widespread branching roots. The alternate leaves have five to nine linear-oblong, dotted leaflets. The leaflets are broader (more than 2 mm wide) than those of Purple Prairie Clover. The terminal inflorescence consists of one to a few spikes (2.5-5 cm long) of crowded, small, white flowers.

Petalostemum purpureum (Vent.) Rydb., PURPLE PRAIRIE CLOVER

Height: 0.3-0.6 meter (1-2 feet)
Flowers: July
Color: Purple
Habitat: Mesic

Identification Features:

The slender, erect, wiry stems of Purple Prairie Clover arise from a short vertical taproot that is highly branched. The alternate leaves have three to five leaflets which are much narrower (less than 2 mm wide) and closer together than those of White Prairie Clover. The inflorescence is a firm cylindric spike of crowded, small, purple flowers with five protruding bright orange-yellow anthers. Purple Prairie Clover is a desirable plant wherever it grows.

Psoralea tenuiflora Pursh, SCURFY PEA

Height: 0.5-1 meter (1.5-3 feet)
Flowers: Mid June-July
Color: Lavender
Habitat: Xeric

Identification Features:

Scurfy Pea has a highly branched, grayish-pubescent, slender stem that arises from a stout taproot. Its alternate leaves are on long petioles and have three to five linear-oblong leaflets with glandular dots. The inflorescences of small (3 mm) lavender flowers are on racemes up to 6 cm long.

Ecological Notes:

Scurfy Pea, like Lead Plant, can survive extreme drought conditions because of anatomical features such as small and gray-colored leaflets which reduce transpiration. Many insects infest this legume including a weevil, *Apion* sp., which destroys its flower bud. An abscission layer at the stem base breaks during late summer allowing the stem and seeds of Scurfy Pea to disperse in a tumbleweed manner.

GENTIANACEAE (GENTIAN)

Gentiana andrewsii Griseb., BOTTLE GENTIAN

Height: 0.2-0.6 meter (0.5-2 feet)
Flowers: September-October
Color: Dark-Light Blue
Habitat: Hydric-Mesic

Identification Features:

Bottle Gentian has a relatively smooth stem that arises from a short rootstock. There are long internodes between the opposite pairs of smooth, parallel-veined leaves. The lance-shaped leaves do not have petioles and tend to bend downward. Tufts of flowers are clustered in the upper leaf axils. The united corolla is nearly closed or "bottle-shaped." The seeds are small, white-cream colored, and winged.

Ecological Notes:

The bumblebee is one of the few insects strong enough to open the bottle-shaped flower and achieve pollination. Attracted by the blue color, the bumblebee presses on the tip of the united corolla and pushes its front half into the "bottle." The entrance behind the bumblebee is held open with its abdomen and rear legs so that the bee does not become trapped.

Gentiana flavida Gray, YELLOWISH GENTIAN

Height: 0.3-0.6 meter (1-2 feet)
Flowers: Mid August-September
Color: Yellowish-Cream
Habitat: Xeric-Mesic

Identification Features:

The stem of Yellowish Gentian is smooth. The internodes between the pairs of leaves are shorter than those of Bottle Gentian; thus the foliage is more clustered than that of Bottle Gentian. Tufts of yellowish-cream flowers are clustered in the upper leaf axils.

Ecological Notes:

Yellowish Gentian thrives quite well in both undisturbed and disturbed prairie.

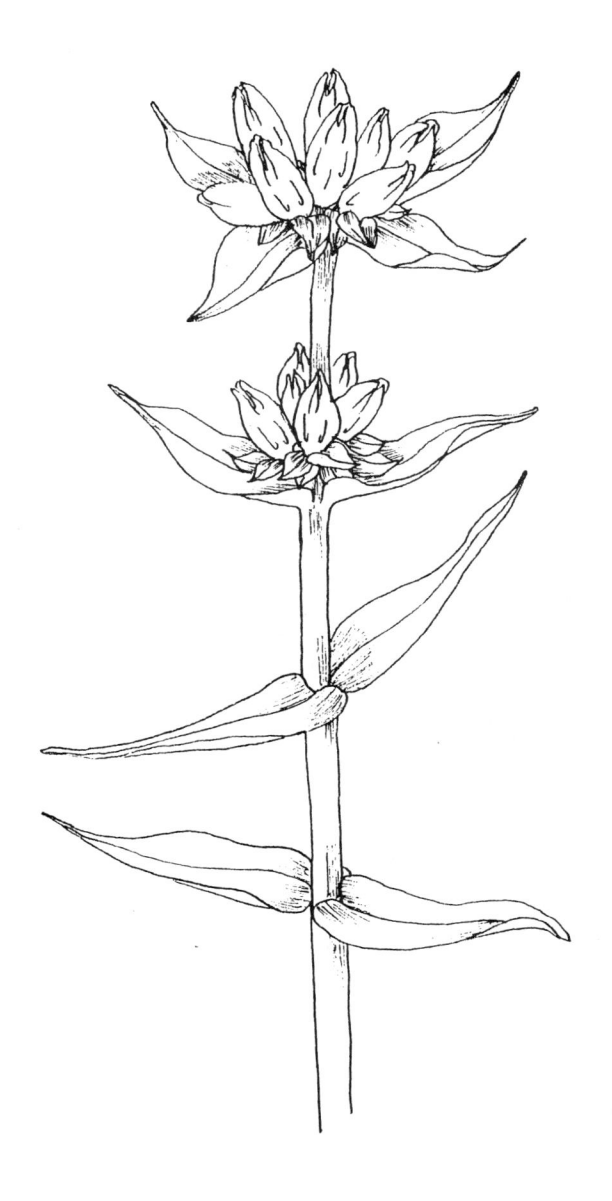

IRIDACEAE (IRIS)

Sisyrinchium albidum Raf., BLUE-EYED GRASS

Height: 0.1-0.25 meter (4-10 inches)
Flowers: Mid May-June
Color: Blue-White
Habitat: Xeric-Mesic-Hydric

Identification Features:

The slender, wiry flattened stems and "grass-like" foliage of Blue-eyed Grass arise from a rootstock. When mature, this species grows in tufts. The "blue-eyes" refer to the flowers which seem to be unnaturally attached to the stems near the tips. Although normally blue, Blue-eyed Grass can be white-flowered in alkaline soil.

Ecological Notes:

Blue-eyed Grass thrives best in stable prairies but can survive in disturbed prairie. The underground rootstocks may be eaten by herbivores that "root" for food.

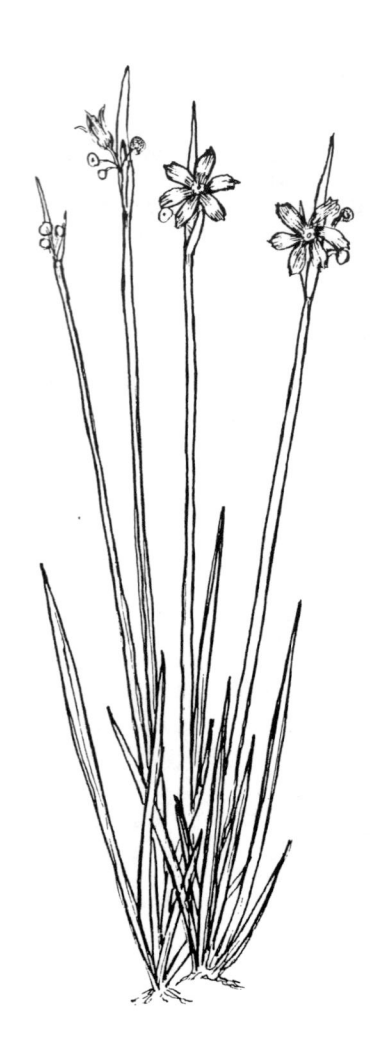

LAMIACEAE (MINT)

Monarda fistulosa L., WILD BERGAMOT (BEEBALM)

Height: 0.6-1.2 meters (2-4 feet)
Flowers: July-August
Color: Lavender
Habitat: Hydric-Mesic-Xeric

Identification Features:

The stem of Wild Bergamot is square, slightly hairy, and arises from a branched rhizome. Its leaves are opposite, ovate-lanceolate shaped, and sharply serrated along their margins. The inflorescences are solitary, terminal heads that have many clusters of two-lipped tubular flowers. Both the leaves and flowers are dotted with glands that secrete volatile, aromatic bergamot oils.

Ecological Notes:

Wild Bergamot often grows in large colonies in prairie, disturbed habitats, and along forest edges. It is pollinated by insects, primarily bumblebees, honeybees, and wasps. The Clear-wing Sphinx Moth also visits this plant. Pollination in Wild Bergamot is interesting due to two major observations. First, inflorescences are composed of several clusters, each with ten or more flowers open at a time, with some flowers in the staminate and others in the pistillate phase. However, young stigmas have a delayed receptivity to accepting self-pollen. Second, bumblebees and honeybees appear to visit staminate and pistillate phase flowers indiscriminately for nectar. Successful cross-pollination and outbreeding of Wild Bergamot is due, at least in part, to the continuous opening of the flowers during the day and the stigma's receptivity to cross-pollen prior to self-pollen.

Pycnanthemum virginianum (L.) Duran & Jackson, COMMON MOUNTAIN MINT

Height: 0.6-0.9 meter (2-3 feet)
Flowers: July-August
Color: White
Habitat: Hydric-Mesic

Identification Features:

The stem of Common Mountain Mint is four-angled or "square." Its narrow leaves are opposite, numerous, and lance-shaped. The white-purplish flowers are in head-like clusters about 5 mm in diameter. Common Mountain Mint is the most aromatic plant of the prairie. When crushed, its plant parts have a strong "mint-like" odor.

Ecological Notes:

In addition to inhabiting hydric-mesic and sometimes xeric prairies, Common Mountain Mint also occurs in habitats with considerable degradation. Its rhizomes extend a few inches from the parent plant and produce new shoots. These new shoots form tight colonies which appear as clumps. Many species of small insects are attracted to this plant.

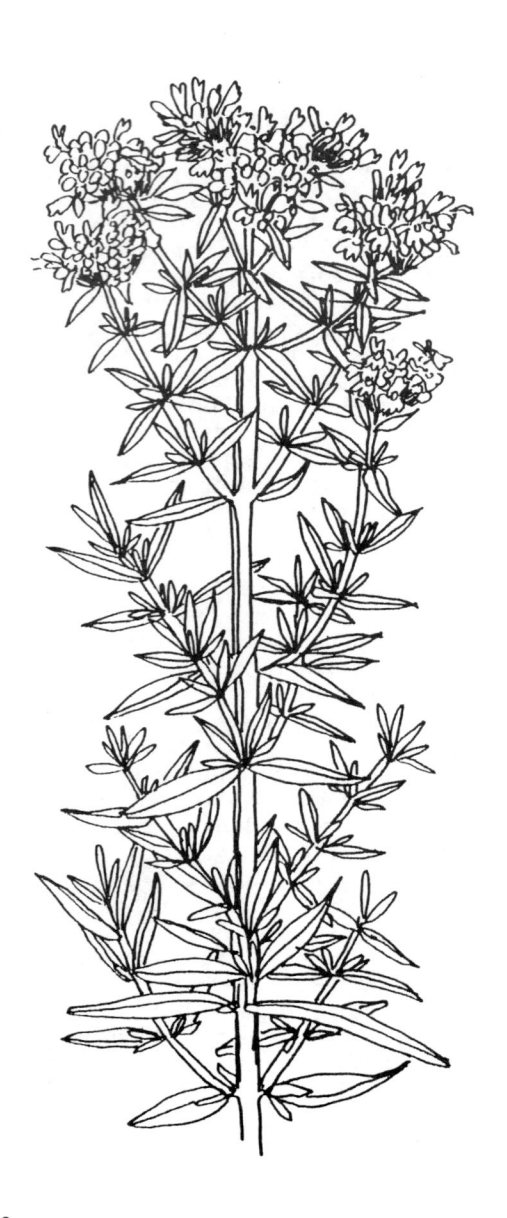

LILIACEAE (LILY)

Allium cernuum Roth, NODDING WILD ONION

Height: 0.3-0.5 meter (1-1.6 feet)
Flowers: August
Color: White-Pink-Lavender
Habitat: Hydric-Mesic

Identification Features:

The stems and leaves of Nodding Wild Onion arise from underground bulbs which are smaller than domestic onions. The leaves are usually long, narrow, tubular, and may be either hollow or solid. White-pink-purple flowers are borne at the tip of leafless stalks in solitary rounded clusters that "nod." When young, the flowering stalk bends in the middle. This bend or "nod" grows toward the top of the stem as the plant matures. The seeds of Nodding Wild Onion are black and hard. When crushed, the plant parts have the distinctive onion odor.

Ecological Notes:

Nodding Wild Onion prefers calcareous soils. Its succulent herbage is grazed by ungulates and its bulbs are dug up by herbivores that "root" for food.

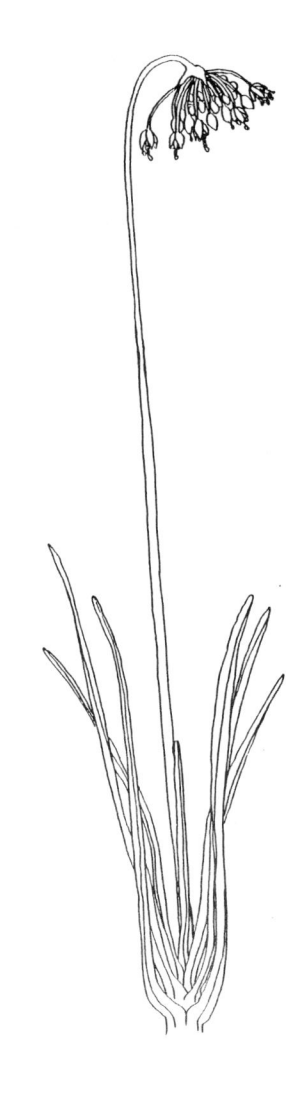

Hypoxis hirsuta (L.) Coville, YELLOW STAR GRASS

Height: 0.05-0.1 meter (2-4 inches)
Flowers: Late May-Early June
Color: Yellow
Habitat: Mesic-Hydric

Identification Features:

The slender floral stem of Yellow Star Grass grows from a tuft of basal, "grass-like" leaves which arise from a small, hairy corm. One to several flowers arise from each slender stalk. Each flower has three yellow petals, three yellow sepals, and blooms for about a day. Yellow Star Grass is a delicate, charming plant.

Ecological Notes:

Yellow Star Grass grows in woods, moist calcareous meadows, and mesic-hydric prairies.

ORCHIDACEAE (ORCHID)

Cypripedium candidum Muhl., WHITE LADY'S SLIPPER

Height: 0.15-0.3 meter (6-12 inches)
Flowers: May
Color: White
Habitat: Hydric

Identification Features:

The stiff, green stalks of White Lady's Slipper are clasped by hairy leaves that have parallel veins. Its leaves ascend up the stem. The small, waxy white "slipper" has green-purple sepals and petals extending outward and over it. The "slipper" has purple veins. Inside the white pouch there are purple speckles. A beauty to behold!

Ecological Notes:

White Lady's Slipper thrives best in moist, alkaline black-soil prairies. There are symbiotic relationships between its roots and soil fungi. Most pollination is accomplished by tiny bees (six to seven mm long) that are attracted to the fragrant, showy lip of this flower. Once inside it may take the small bee, or an occasional wasp or beetle, up to 15 minutes to crawl out of the curved, smooth lip walls of this beautiful blossom. While exiting, the bee gets smeared with sticky, green pollen and carries it to the next blossom. Since the blossom lacks nectar, the insect receives no nourishment for its efforts.

POACEAE (GRASS)

General importance of the grass family to the prairie and other notes:

The grasses contribute most of the prairie plant biomass and provide fuel for prairie fires. Humus from the decay of their deep and extensive root systems is mainly responsible for the rich, black color of grassland soils.

The growth habit of prairie grasses is sod-forming or bunch-forming, or both. Sod-forming grasses, such as cordgrass and switch grass, reproduce sexually from seed and asexually from underground stems called rhizomes. The rhizomes extend horizontally from a few inches to a few feet from the parent plant and produce new shoots from their tips or from the nodes along the stem. This results in a dense stand of shoots that may completely occupy the soil. A few sod-forming grasses also reproduce from horizontal stems above the ground called stolons.

Bunch-forming grasses have an erect growth of many shoots at their base. These shoots are called tillers. Little bluestem and prairie dropseed are bunch-forming grasses. Like sod-forming grasses, bunch-forming grasses reproduce sexually and asexually. Sometimes a grass species, such as big bluestem, may have both sod-forming and bunch-forming growth habits.

Besides the taxonomic classification, prairie grasses may be grouped according to their climatic origin as either "cool-season" or "warm-season" grasses. Cool-season grasses, such as needlegrass and bluejoint, are of northern origin. They renew their growth in early spring and usually mature from April to early July. Cool-season grasses become semi-dormant during the hot summer months and renew growth during autumn. By contrast, warm-season grasses, such as big and little bluestem, begin growth during late spring and grow continuously during the summer and into early fall. They mature during mid-fall. Warm-season and cool-season grasses mingle freely as their requirements overlap.

Grasses have several anatomical features which enable them to flourish in the relatively harsh environment of the prairie. Some of these anatomical features include:

(1) deep root systems that reach depths of 2-4 m and branch profusely, enabling the grasses to obtain water during periods of drought,

(2) tough stems, reinforced by a silicon oxide, which enable the grasses to withstand high winds, and

(3) specialized "hinge cells" in the upper epidermis of the leaf. During periods of drought they lose water rapidly, then contract and cause the leaf to roll up in a long tube. The pores through which water vapor is normally transpired are now inside the tube, while the exposed lower surface of the leaf is highly resistant to water loss.

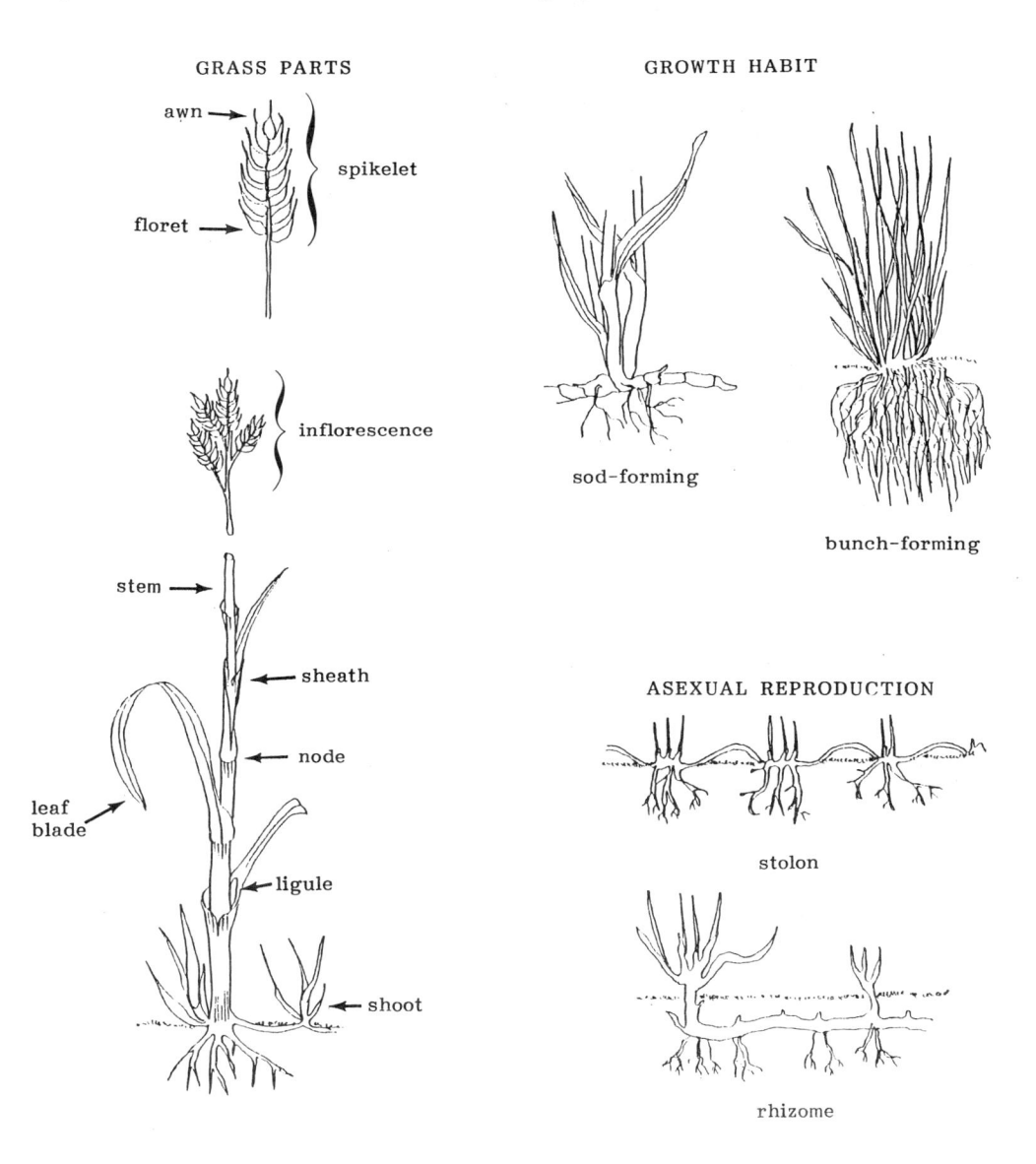

GRASS PARTS

awn
spikelet
floret

inflorescence

stem
sheath
node
leaf blade
ligule
shoot

GROWTH HABIT

sod-forming

bunch-forming

ASEXUAL REPRODUCTION

stolon

rhizome

Andropogon gerardi Vitman, BIG BLUESTEM (TURKEY FOOT)

Height: 1-3 meters (4-10 feet)
Flowers: August
Color: Yellow anthers
Habitat: Mesic-Xeric-Hydric

Identification Features:

Big Bluestem has both sod-forming and bunch-forming growth habits. The leaves of Big Bluestem remain green for most of the summer. During late summer and early fall its wax-like stem coating turns bluish-purple, and following a frost the foliage turns reddish-purple. Big Bluestem takes its popular common name from the bluish-purple color of the stem and nodes; however, it is also called "turkey foot" because its seed spikes frequently branch into three parts. Its numerous leaves may be up to 1.2 cm wide and are often hairy. The yellow anthers of its flowers are rather showy.

Ecological Notes:

This warm-season grass is one of the most widely spread and important dominants of the tall grass prairie. Reasons for its dominance include rapid growth, dense sod-forming habit, height, and shade tolerance of mature plants and seedlings. In lowland tall grass communities, Big Bluestem often grows in nearly pure stands.

Big Bluestem has a high protein content and is highly palatable for both native and introduced ungulates. Its deep fibrous root system reaches depths of 1.5 to 2.2 m and branches profusely. Humus from decay of its root system is very important in soil formation, and greatly contributes to the rich, black color of grassland soil. Big Bluestem provides much fuel for prairie fires.

Andropogon scoparius Michx., LITTLE BLUESTEM

Height: 0.5-1.1 meters (1.5-3.2 feet)
Flowers: Late August-September
Color: Pinkish-White seed spikes
Habitat: Xeric-Mesic

Identification Features:

Little Bluestem is a strong bunch-grass. The basal shoots and stem nodes are bluish-purple. The leaves are more slender and wiry than those of Big Bluestem. Other features distinguishing Little Bluestem from Big Bluestem are a somewhat flattened basal portion of stems and leaf sheaths, its slightly folded leaves, and its lack of hairiness on both the sheaths and lower portions of leaves. During fall, the pinkish colored stems have fuzzy and fluffy white-silvery seed spikes.

Some botanists classify this plant as *Schizachyrium scoparium* Nash.

Ecological Notes:

This warm-season grass grows in many habitats with soils ranging from deep to shallow and rocky, and sandy to clay textured. It is an upland dominant but intermingles with tall grasses on lower sites. The roots reach depths of 1.4-1.8 m, are very abundant, and form a dense sod. Little Bluestem is a keen competitor as it possesses a deep and extensive root system along with a reduced transpiring leaf blade surface. These characteristics enable Little Bluestem to survive periods of drought.

Bouteloua curtipendula (Michx.) Torr., SIDE-OATS GRAMA

Height: 0.4-1 meter (1.5-3 feet)
Flowers: Late July-August
Color: Crimson-Red anthers
Habitat: Xeric

Identification Features:

Side-oats Grama takes its name from the oat-like florets which appear to hang from one side of the seed stalk. It retains the zigzag-topped seed stalks into autumn. The leaves are normally flat and have stiff hairs along the bases of their edges. While maturing, the tips of the upper leaves die as the basal leaves curl and turn light colored. The flowers have beautiful, showy, crimson-red stamens.

Ecological Notes:

Side-oats Grama is a warm-season, sod-forming grass which prefers xeric upland communities with alkaline conditions. Its dense and well-branched root system, extending to 0.5 m deep, can occupy deep or shallow soil. Side-oats Grama does not tolerate shade from the taller grasses. It is a weak grass competitor species unless the area is heavily grazed or extremely dry. Under these conditions, its short rhizomes promote the formation of small areas of sod.

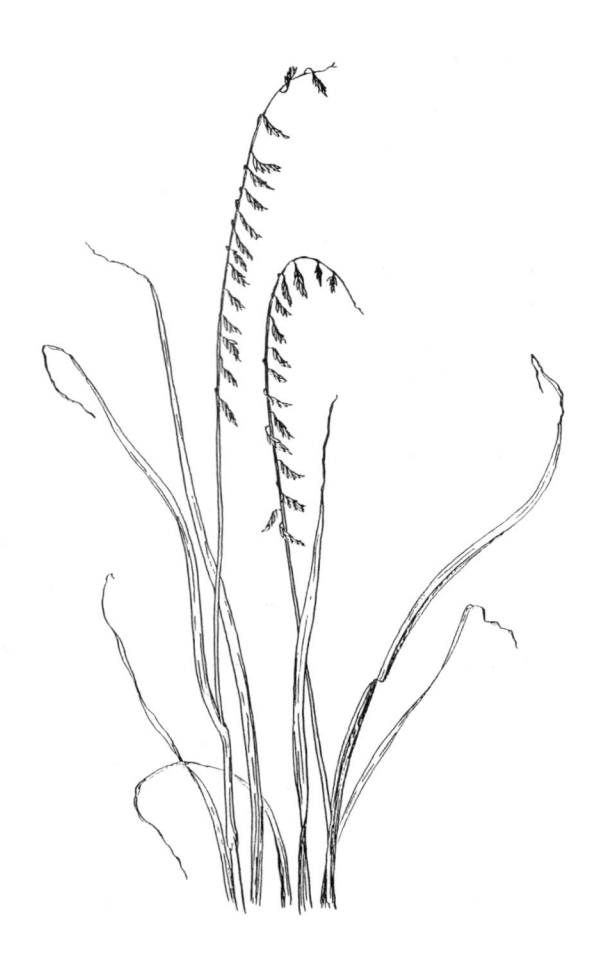

Bromus kalmii Gray, PRAIRIE BROME

Height: 0.3-0.7 meter (1-2 feet)
Flowers: June-July
Color: Inconspicuous
Habitat: Mesic

Identification Features:

Prairie Brome is a solitary grass with soft, drooping inflorescences that are somewhat open and spreading. The leaf blades are flat and the edges of the sheath grow together to form a tube.

Ecological Notes:

This cool-season grass thrives best in calcareous soils.

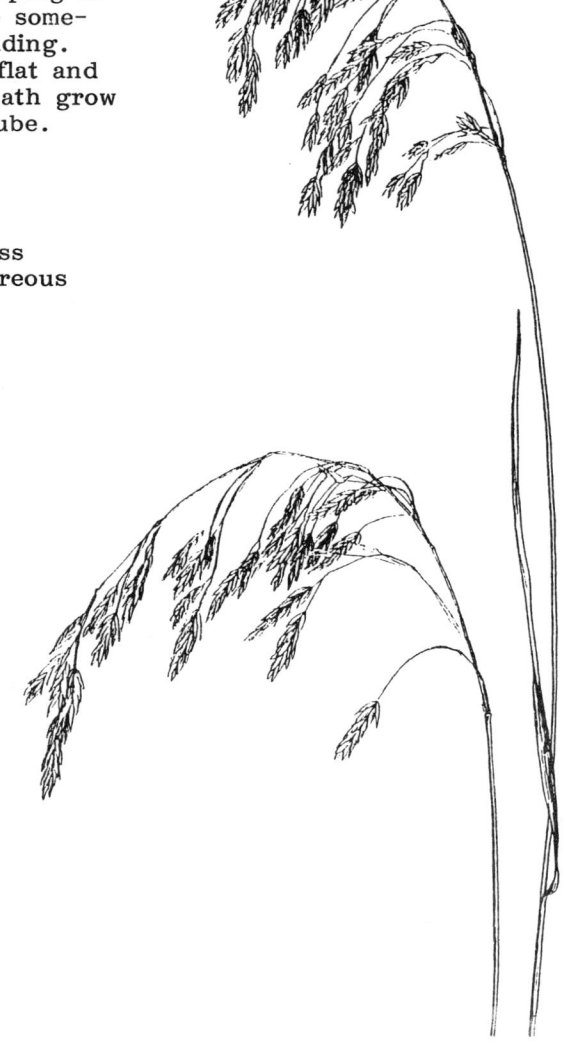

Calamagrostis canadensis (Michx.) Nutt., BLUE JOINT GRASS

Height: 0.6-1.2 meters (2-4 feet)
Flowers: June-Early July
Color: Inconspicuous
Habitat: Hydric-Mesic

Identification Features:

Blue Joint has long beautiful blue-green leaf blades and grows in dense clumps. The inflorescences are approximately 15 cm long and have white, silky hair on one side. The nodes between leaf blades are usually swollen.

Ecological Notes:

This cool-season grass often forms solid stands eliminating other species, especially in marshy sites. Blue Joint is nutritious and relished by ungulates.

Hierochloe odorata (L.) Beauv., VANILLA GRASS

Height: 0.3-0.6 meter (1-2 feet)
Flowers: May
Color: Amber spikelets
Habitat: Hydric-Mesic

Identification Features:

This fragrant, vanilla-scented grass forms dense sod clumps. The inflorescences are 7-8 cm long and to one side. The leaf blades are small while the plant is in bloom; however, long blades grow later.

Ecological Notes:

This cool-season grass thrives best in calcareous, moist soils.

Panicum virgatum L., SWITCH GRASS

Height: 1-2 meters (4-6.2 feet)
Flowers: August
Color: Reddish-Purple anthers
Habitat: Mesic-Xeric-Hydric

Identification Features:

Switch Grass is a sod-forming grass. It has strong glossy leaves. Identification is simplified by the somewhat inverted V-shaped wedge of fine, wavy hairs along the leaf's upper surface where its leaf blade joins the sheath. The large, open panicles are up to 40 cm long and begin to unfold in July. Its small inconspicuous flowers have reddish-purple stamens. Large tear-drop shaped seeds are borne on the open panicles.

Ecological Notes:

Switch Grass is a warm-season species that thrives in a wide range of habitats. Its roots are 3-4 mm in diameter and penetrate almost vertically to depths of 2 to 3.7 m. Switch Grass is aggressive, often forming dense stands and excluding other species.

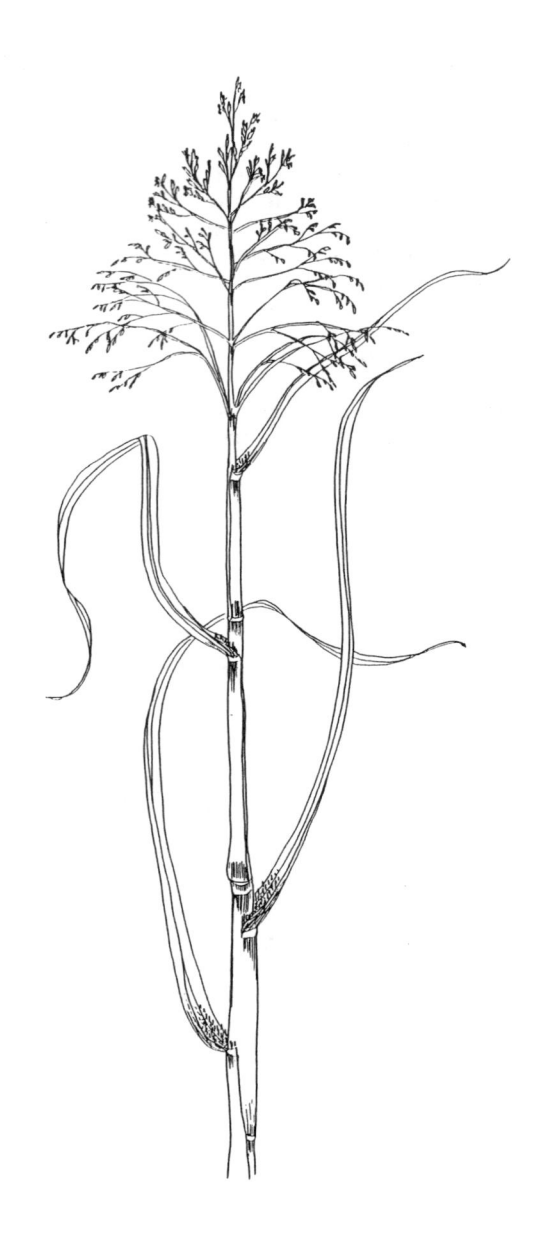

Sorghastrum nutans (L.) Nash, INDIAN GRASS

Height: 1-2 meters (3-7 feet)
Flowers: August-September
Color: Yellow anthers
Habitat: Mesic-Xeric-Hydric

Identification Features:

The leaves of Indian Grass are rather stiff and spread at a 45 degree angle to the stem. They are lighter green and stiffer than those of Big Bluestem. The upper surface of the leaf blade possesses a prominent claw-like ligule where it joins the sheath. Showy yellow anthers protrude when the plant is in bloom. During fall the soft, silky-textured plumes become a beautiful golden color and gently sway with a breeze.

Ecological Notes:

This warm-season grass is a common associate of Big Bluestem. Both species have approximately the same growth habits, nutrient and moisture requirements. The root system of Indian Grass reaches depths of 1.6-1.8 m, extends laterally, and is slowly rhizomatous in the absence of tough competition. The decay of its deep, lateral fibrous root system is very important in soil formation. Indian Grass is relished by native and domestic ungulates. It also provides much fuel for prairie fires.

Spartina pectinata Link, PRAIRIE CORDGRASS (SLOUGHGRASS)

Height: 1-2.5 meters (3-8 feet)
Flowers: July-August
Color: Yellow anthers
Habitat: Hydric

Identification Features:

The name "cordgrass" is suggestive of the toughness of its long, coarse leaves (up to 80 cm) and thick, tough stems. The leaf blades have finely serrated (saw-toothed) margins that are quite sharp and capable of cutting skin. Showy yellow anthers protrude from the blooming florets. The seed spikes are up to 6 cm long. Cordgrass turns golden-yellow in the fall.

Ecological Notes:

This warm-season grass species occupies soils too wet and too poorly drained for the development of Big Bluestem and Switch Grass communities. Roots can penetrate to depths of 2.5-4 m into water-logged (poorly aerated) soil due to the large air-conducting space in its cortex. The upper 15-25 cm of soil beneath stands of Cordgrass is occupied by a mat of course, thick (5-10 mm), almost woody, highly branched rhizomes. This sod-forming grass is aggressive in hydric soil conditions, often forming dense stands and excluding other species. In the center of solid stands of Cordgrass, grass height is much lower and there is rarely any flowering. Flowering usually occurs in plants around the edges. Plants in the center of dense stands store starch in their rhizomes to ensure survival rather than produce flowers or height. Cordgrass is relished by ungulates during spring and early summer before it becomes too tough.

Sporobolus heterolepis Gray, PRAIRIE DROPSEED

Height: 0.6-1.2 meters (2-4 feet)
Flowers: August
Color: Purplish stamens
Habitat: Mesic-Xeric

Identification Features:

Prairie Dropseed is a most attractive bunch-grass that grows in fountain-like tufts of 10-20 cm in diameter. Its long, slender leaves (60 or more cm) have a smooth texture. The inflorescence is a well-developed panicle about 20 cm long. Following a burn, Prairie Dropseed bunches are noticeable as firm little mounds on the prairie soil.

Ecological Notes:

The presence of this warm-season grass in a prairie remnant indicates virgin conditions, as Prairie Dropseed does not tolerate heavy grazing or soil disturbances such as plowing. Its fibrous roots reach a depth of 1-1.7 m. The decay of these roots helps build the rich fertile soil of the grasslands. In some upland communities, it can be the dominant grass species. Prairie Dropseed is relished by ungulates and also provides fuel for prairie fires.

Stipa spartea Trin., NEEDLE (PORCUPINE) GRASS

Height: 0.5-1 meter (1.5-3 feet)
Flowers: Mid May-June
Color: Inconspicuous
Habitat: Xeric

Identification Features:

Needle Grass gets its common name from the sharp, long twisting awns. This bunchgrass has long tapering leaves which are corrugated on the upper surface but smooth and shining beneath. Unlike other prairie grasses that turn a golden yellow or tan when mature, its leaves become nearly white.

Ecological Notes:

This cool-season bunchgrass has strong, fibrous roots (1 to 1.5 mm) that reach depths of 0.7-1 m. During periods of drought, the leaves of Needle Grass roll upward and inward and form a tube. The pores through which water vapor is normally transpired are now inside the tube while the exposed lower surface of the leaf is highly resistant to water loss. Upon falling to the ground or coming into contact with an object, the sharp point of the seed penetrates by the twisting action of its awn. Needle Grass is very nutritious and relished by ungulates. However, it can also cause mechanical injury to livestock as the sharp point of the seed screws through fur and into the flesh by the twisting action of its awn.

POLEMONIACEAE (PHLOX)

Phlox glaberrima interior Wherry, MARSH PHLOX

Height: 0.3-0.9 meter (1-3 feet)
Flowers: July-August
Color: Violet-Pink
Habitat: Hydric-Mesic

Identification Features:

The stem of Marsh Phlox is slender, erect, and smooth. Its opposite leaves are sessile, stiff, narrow and tapered. The inflorescence is a group of loosely branched clusters.

Ecological Notes:

Marsh Phlox thrives best in moist, calcareous prairies. It is pollinated by hummingbirds and long-tongued butterflies.

Phlox pilosa L., PRAIRIE PHLOX

Height: 0.3-0.7 meter (1-2 feet)
Flowers: Mid May-June
Color: Pink-Lavender
Habitat: Xeric-Mesic

Identification Features:

The stems of Prairie Phlox are slender, spreading, and hairy. Its opposite leaves are sessile, stiff, and slightly broader than those of Marsh Phlox. Large clusters of tight buds are furled in the shape of an umbrella. These clusters of buds develop into umbrella-like inflorescences of beautiful pink-lavender flowers. The flowers are on short pedicels and have a tube-like calyx.

Ecological Notes:

Prairie Phlox is often an indicator of virgin prairie.

PRIMULACEAE (PRIMROSE)

Dodecatheon meadia L., SHOOTING STAR

Height: 0.2-0.6 meter (9-24 inches)
Flowers: May-Early June
Color: White-Pink-Lavender
Habitat: Xeric-Mesic-Hydric

Identification Features:

One or more stiff, green stems of Shooting Star arise from a rosette of long (7-20 cm), strap-like, pale green basal leaves with reddish-pink midribs. The flower clusters from these upright stems are not crowded, each flower being on a long stalk. The flowers are uniquely shaped with their five recurved petals, protruding pistil, and tightly pressed stamens pointing out. By July, the leaves have disappeared and only the stalk with its seed capsule remains.

Ecological Notes:

Shooting Star grows in prairie, savanna, rocky hillsides and moist slopes. Pollination is by bumblebees.

SEED CAPSULES

RANUNCULACEAE (BUTTERCUP)

Anemone cylindrica Gray, THIMBLEWEED

Height: 0.3-0.7 meter (1-2 feet)
Flowers: June-July
Color: Greenish-White
Habitat: Mesic-Xeric

Identification Features:

The tall, slender, fibrous stem of Thimbleweed arises from a bulbous rhizome. It has a whorl of deeply cut five-parted basal leaves with long petioles. Flowers arise from the tip of the top whorl of smaller leaves. The cylindrical receptacle has many pistils and stamens surrounding it. This receptacle is "thimble-shaped," thus giving this plant its common name. During fall, the thimble slowly disintegrates and its cottony seeds disperse into the air.

Ecological Notes:

Of all the Anemones, Thimbleweed has the greatest fidelity to prairie. The cottony seeds and fluff which remain until spring are sometimes gathered by hummingbirds and used to construct their minitature nests.

RHAMNACEAE (BUCKTHORN)

Ceanothus americanus L., NEW JERSEY TEA

Height: 0.3-1 meter (1-3 feet)
Flowers: Late June-July
Color: White
Habitat: Mesic

Identification Features:

This low, upright shrub has numerous slender stems arising from huge rootstocks. Clusters of white flowers arise from the tops of these stems. The dark green leaves of New Jersey Tea are alternate, finely toothed, veiny, oval, and stiff, with fine hairs beneath. As the seed capsules mature, they become black and are easily recognized. New Jersey Tea is a most attractive shrub at all stages during the growing season.

Ecological Notes:

Despite its common name, New Jersey Tea is found growing in prairies and savannas. The flowers are fragrant and attract many species of butterflies and other insects. Herbivores, especially deer and rabbits, browse heavily upon this shrub. New Jersey Tea can be burned or cut off to the ground while dormant, but will flower by early July on the new wood.

ROSACEAE (ROSE)

Potentilla arguta Pursh, PRAIRIE CINQUEFOIL

Height: 0.5-0.8 meter (1.5-2.5 feet)
Flowers: Late June-July
Color: White
Habitat: Mesic-Xeric

Identification Features:

The stout stem of Prairie Cinquefoil is highly pubescent. Its alternate leaves usually have seven to eleven pubescent leaflets. The leaflets have sharp coarse teeth along their margins. The flowers are in a rather dense cluster and resemble those of a strawberry. The tiny seeds of Prairie Cinquefoil are dry and enclosed within the infolded sepals.

Ecological Notes:

Prairie Cinquefoil is a member of the stable prairie and does not tolerate much disturbance.

SAXIFRAGACEAE (SAXIFRAGE)

Heuchera richardsonii grayana R. Br., PRAIRIE ALUM ROOT

Height: 0.3-1 meter (1-3 feet)
Flowers: June-Early July
Color: Orange-Creamish-Brown-Green
Habitat: Xeric-Mesic-Hydric

Identification Features:

The flowering stem of Prairie Alum Root arises from a cluster of geranium-like leaves. This broad cluster of scalloped, round, hairy leaves is handsome throughout the growing season. The flowers are small (3-5 mm) and bell-like, with beautiful orange anthers protruding out on green filaments.

Ecological Notes:

Aphids often attach themselves to the small attractive flowers.

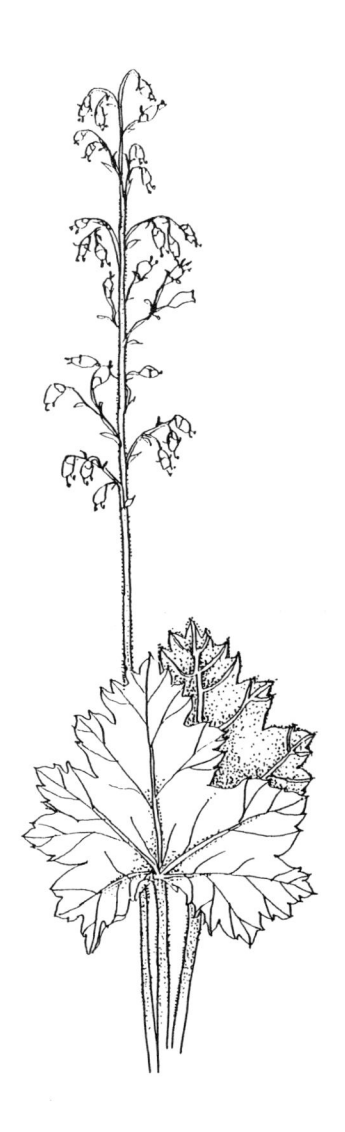

SCROPHULARIACEAE (FIGWORT)

Veronicastrum virginicum (L.) Farw., CULVER'S ROOT

Height: 0.6-1.7 meters (2-5 feet)
Flowers: July-August
Color: White
Habitat: Hydric-Mesic

Identification Features:

The tall, erect stems of Culver's Root have leaves whorled around each of their nodes. Its leaves are 7-20 cm long and 2-2.5 cm wide and have slightly saw-toothed edges. The tiny, white flowers (about 2 mm long) are densely crowded on several spike-like racemes that are up to 20 cm long. The racemes may have several branches.

Ecological Notes:

The presence of Culver's Root indicates good soil and moisture conditions. It has a fragrant scent and is visited by many insects such as small beetles and sulfur butterflies.

SELECTED REFERENCES

Cruden, R. W., L. Hermanutz, and J. Shuttleworth. 1984. The pollination biology and breeding system of *Monarda fistulosa* (Labiatae). Oecologia 64: 104-110.

Frost, S. W. 1945. Insects feeding or breeding on indigo, *Baptisia*. *Journal New York Entomological Society*, 53: 219-225.

Gleason, Henry Allan. 1974. *The New Britton and Brown Illustrated Flora of the Northeastern United States and Adjacent Canada.* 5th printing. New York: New York Botanical Garden.

Haddock, R. C. and S. J. Chaplin. 1982. Pollination and seed production in two phenologically divergent prairie legumes (*Baptisia leucophaea* and *B. leucantha*). *The American Midland Naturalist*, 108: 175-186.

Johnson, James R. and James T. Nichols. 1970. *Plants of South Dakota Grasslands*, South Dakota State University, Brookings, S. D.

Kerster, H. W. 1968. Population age structure in the prairie forb, *Liatris aspera*. Bioscience 18(5): 430-432.

Madson, John. 1982. *Where the Sky Began: Land of the Tall Grass Prairie*. Sierra Club Books, San Francisco.

McKone, Mark J. and David D. Biesboer. 1986. Nitrogen fixation in association with the root systems of goldenrods (*Solidago* L.). Soil Biol. Biochem., 18(5): 543-545.

Owensby, Clenton E. 1980. *Kansas Prairie Wildflowers*. Iowa State University Press, Ames, Iowa.

Runkel, Sylvan T. and Dean M. Roosa. 1989. *Wildflowers of the Tall Grass Prairie: The Upper Midwest*. Iowa State University Press, Ames, Iowa.

Schaal, Barbara A. 1978. Density dependent foraging on *Liatris pycnostachya*. Evolution 32(2): 452-454.

Swink, Floyd and Gerould Wilhelm. 1979. *Plants of the Chicago Region*, The Morton Arboretum, Lisle, Illinois.

Voight, John W. and Robert H. Mohlenbrock. 1978. *Prairie Plants of Illinois*, Department of Conservation, Springfield, Illinois.

Voss, John and Virginia S. Eifert. 1951. *Illinois Wild Flowers,* Department of Registration and Education, Springfield, Illinois.

Weaver, John E. 1954. *North American Prairie*, Johnsen Publishing Company, Lincoln, Nebraska.

Werner, Patricia A. 1978. On the determination of age in *Liatris aspera* using cross-sections of corms: Implications for past demographic studies. *The American Naturalist*, 112(988): 1113-1119.

GLOSSARY

Abscission layer. A layer of special cells at the base of a plant part, such as the stem, that allows it to separate from the rest of the plant.

Achene. A small, dry, hard, one-seeded fruit.

Anatomical. Referring to the structural make-up of an organism or any of its parts.

Anther. The pollen-bearing part of the stamen.

Awn. A bristle-like appendage, common on grasses, attached to a lemma or glume.

Axil. The angle between two plant parts, such as the stem and leaf.

Blade. The flattened expanded portion of a leaf.

Biomass. The amount of organic matter per unit area or volume.

Bract. A reduced or otherwise modified leaf which subtends a flower or flower cluster.

Bulb. Usually an underground leaf bud, consisting of a thickened short stem that is overlapped with scalelike leaves, as in lily or onion.

Calcareous. Of or pertaining to a high concentration of calcium carbonate (limestone) in the soil.

Calyx. The outer whorl (usually green) of flower parts; collective for sepals.

Corm. A solid bulblike stem that is usually underground.

Corolla. The inner whorl (usually colored) of flower parts; collective term for the petals.

Cylindric. Elongated with a circular cross section.

Epidermis. The covering tissue of roots and leaves, and stems of non-woody plants.

Filament. The threadlike stalk of the stamen, which supports the anther.

Floret. A very small flower, particularly one found in a dense inflorescence, as in the composites or in the grasses.

Forb. A herbaceous plant other than grass, sedge, or rush.

Fungi. Organisms, comprised of molds, mildews, rusts, smuts, and mushrooms, that lack chlorophyll and reproduce mainly by means of asexual spores.

Glume. One of the two empty chafflike bracts at the base of a grass spikelet.

Herbivore. An organism that eats plants and plant products.

Hydric. Wet, especially relating to the soil moisture conditions.

Inflorescence. A flower cluster.

Internode. The portion of the stem between two successive nodes.

Involucral. Referring to the involucre which are whorls of bracts subtending an inflorescence; as around the base of a composite head.

Lanceolate. Lance-shaped; long tapering above the middle and several times longer than wide.

Lemma. The lower of two scale-like bracts that enclose a grass flower or seed; located directly above the glumes.

Lignified. "Woody" as a result of conversion of compounds of the cell wall into lignin.

Ligule. A small strap-shaped appendage at the junction of the blade and sheath of a grass leaf; also, one of the strap-shaped corollas of the composites.

Loment. A flat legume fruit that is conspicuously constricted between the seeds, falling apart at the constrictions when mature into one-seeded joints.

Mesic. Medium, especially relating to the soil moisture conditions.

Midrib. The central or main vein of a leaf.

Mucilaginous. Slimy, as sap, composed of gelatinous substances and carbohydrates.

Nectar. A sweet liquid rich in sugars, amino acids, and other compounds that is produced by various plant parts such as the stigma and nectar glands.

Node. The point on a stem where leaves, branches, or inflorescences arise.

Ovate. Egg-shaped, the broadest part below the middle.

Palmate. Having parts deeply lobed and diverging from a common base.

Palatability. A quality of forage that is pleasing and acceptable to the taste of a grazing animal.

Pappus. Outgrowths on the achenes of composites consisting of bristles, hairs, scales or awns.

Panicle. A cluster of flowers arranged in a series of two or more racemes.

Pedicel. The stalk of a single flower in an inflorescence.

Petal. A part of the corolla.

Petiole. The stalk of a leaf.

pH. The relative concentration of H^+ ions in a solution. Low pH values (less than 7) indicate high concentrations of H^+ ions (acidic), and high pH values (greater than 7) indicate low concentrations of H^+ ions (alkaline or basic).

Pistil. The female reproductive structure of a flower consisting of the stigma, style, and ovary.

Predation. The relationship in which one organism (predator) consumes another organism (prey) for its nutrition and survival.

Pubescent. Having the surface covered with soft hairs.

Raceme. An inflorescence whose stalked flowers are arranged along an elongated axis.

Ray. A strap-shaped (ligulate) marginal flower in the head of the composite inflorescence.

Receptacle. The floral axis to which the various flower parts are attached; for example, the disk or dome-shaped structure of the composite family that bears the florets.

Rhizome. An underground, horizontal stem which can produce shoots and roots at the nodes, giving rise to new plants.

Rootstock. An underground stem; used to designate a rhizome or rhizome-like structure.

Rosette. A cluster of leaves in a circular arrangement at the base of a plant.

Sepal. A part of the calyx.

Sessile. Without a stalk of any kind.

Sheath. A tubular structure, consisting of the lower part of the leaf, which clasps or encloses the stem, especially in grasses and sedges.

Spike. An unbranched inflorescence in which the flowers are sessile or sub-sessile on a central axis.

Stamen. The pollen-bearing organ of the flower, composed of the anther and filament.

Stigma. The part of a pistil or style that receives the pollen.

Stipule. A leaf-like appendage found at the point of attachment of a leaf to the stem; usually occur in pairs.

Stolon. A modified, above-ground, horizontal stem that roots at the nodes and produces new plants.

Succulent. Thickened, as leaves, stems, or roots that are juicy and fleshy.

Symbiotic. Living together, referring to two dissimilar organisms, with benefit, to one or both, but without harm to either.

Taproot. The primary descending root, usually thickened and larger than others in the root system.

Transpiration. The loss of water from the plant to the atmosphere, occurring mainly through evaporation at leaf stomata.

Tuber. An enlarged, thickened, fleshy portion of an underground stem, usually functioning as a food reserve organ.

Ungulate. A hoofed mammal, such as a buffalo.

Xeric. Dry, especially relating to soil moisture conditions.

FAMILY NAME INDEX

Apiaceae (Carrot-Parsley), 1

Asclepiadaceae (Milkweed), 3

Asteraceae (Composite-Sunflower), 6

Boraginaceae (Borage), 28

Commelinaceae (Spiderwort), 29

Cyperaceae (Sedge), 30

Euphorbiaceae (Spurge), 31

Fabaceae (Legume), 32

Gentianaceae (Gentian), 44

Iridaceae (Iris), 46

Lamiaceae (Mint), 47

Liliaceae (Lily), 49

Orchidaceae (Orchid), 51

Poaceae (Grass), 52

Polemoniaceae (Phlox), 65

Primulaceae (Primrose), 67

Ranunculaceae (Buttercup), 68

Rhamnaceae (Buckthorn), 69

Rosaceae (Rose), 70

Saxifragaceae (Saxifrage), 71

Scrophulariaceae (Figwort), 72

SCIENTIFIC NAME INDEX

Allium cernuum, 49

Amorpha canescens, 33

Andropogon gerardi, 54

   A. scoparius, 55

Anemone cylindrica, 68

Asclepias sullivantii, 4

   A. tuberosa, 5

Aster, 7

   A. ericoides, 7

   A. laevis, 8

   A. novae-angliae, 9

Baptisia, 34-35

   B. leucantha, 36

   B. leucophaea, 37

Bouteloua curtipendula, 56

Bromus kalmii, 57

Cacalia tuberosa, 10

Calamagrostis canadensis, 58

Carex bicknellii, 30

Ceanothus americanus, 69

Coreopsis palmata, 11

   C. tripteris, 12

Cypripedium candidum, 51

Desmodium, 38

   D. canadense, 38

   D. illinoense, 39

Dodecatheon meadia, 67

Echinacea pallida, 13

Eryngium yuccifolium, 1

Euphorbia corollata, 31

Gentiana andrewsii, 44

   G. flavida, 45

Heuchera richardsonii grayana, 71

Hierochloe odorata, 59

Hypoxis hirsuta, 50

Lespedeza capitata, 40

Liatris, 14

   L. aspera, 15

   L. pycnostachya, 16

Lithospermum canescens, 28

Monarda fistulosa, 47

Panicum virgatum, 60

Parthenium integrifolium, 17

Petalostemum, 41

   P. candidum, 41

   P. purpureum, 42

Phlox glaberrima interior, 65

   P. pilosa, 66

Potentilla arguta, 70

Psoralea tenuiflora, 43

Pycnanthemum virginianum, 48

Ratibida pinnata, 18

Rudbeckia hirta, 19

Schizachyrium scoparium, 55

Senecio pauperculus balsamitae, 20

Silphium, 21

   S. integrifolium, 22

   S. laciniatum, 23

   S. perfoliatum, 24

   S. terebinthinaceum, 25

Sisyrinchium albidum, 46

Solidago, 26

   S. nemoralis, 26

   S. rigida, 27

Sorghastrum nutans, 61

Spartina pectinata, 62

Sporobolus heterolepis, 63

Stipa spartea, 64

Tradescantia ohiensis, 29

Veronicastrum virginicum, 72

Zizia aurea, 2

COMMON NAME INDEX

Alum root, 71
Balsam ragwort, 20
Bicknell's sedge, 30
Big bluestem, 54
Black-eyed Susan, 19
Blue-eyed grass, 46
Blue joint grass, 58
Bottle gentian, 44
Butterfly milkweed, 5
Button blazing star, 15
Canada tick trefoil, 38
Common mountain mint, 48
Compass plant, 23
Cream wild indigo, 37
Culver's root, 72
Cup plant, 24
Flowering spurge, 31
Golden Alexander, 2
Gray-headed coneflower, 18
Heath aster, 7
Illinois tick trefoil, 39
Indian grass, 61
Indian plantain, 10
Lead plant, 33

Little bluestem, 55
Marsh phlox, 65
Mountain mint, 48
Needle grass, 64
New England aster, 9
New Jersey tea, 69
Nodding wild onion, 49
Old-field goldenrod, 26
Porcupine grass, 64
Prairie alum root, 71
Prairie blazing star, 16
Prairie brome, 57
Prairie cinquefoil, 70
Prairie cordgrass, 62
Prairie coreopsis, 11
Prairie dock, 25
Prairie dropseed, 63
Prairie milkweed, 4
Prairie phlox, 66
Prairie puccoon, 28
Prairie sedge, 30
Purple coneflower, 13
Purple prairie clover, 42
Rattlesnake master, 1

Rigid goldenrod, 27
Rough blazing star, 15
Rosinweed, 22
Round-headed bush clover, 40
Scurfy pea, 43
Shooting star, 67
Showy tick trefoil, 38
Side-oats grama, 56
Sloughgrass, 62
Smooth blue aster, 8
Spiderwort, 29
Stiff goldenrod, 27
Switch grass, 60

Tall coreopsis, 12
Thimbleweed, 68
Turkey foot, 54
Vanilla grass, 59
White Lady's slipper, 51
White prairie clover, 41
White wild indigo, 36
Wild bergamot, 47
Wild quinine, 17
Yellow coneflower, 18
Yellowish gentian, 45
Yellow star grass, 50